Dark Sky Dreamings

an Inland Skywriters Anthology

Merrill Findlay, editor

with Suzie Gibson, Val Clark & Gai Lander

Interactive Press
Brisbane

Interactive Press
an imprint of IP (Interactive Publications Pty Ltd)
Treetop Studio • 9 Kuhler Court
Carindale, Queensland, Australia 4152
ipoz.biz/interactive-press
ipoz.biz/ipstore

First published by IP in 2019

Printed in 12 pt Baskerville on 14 pt Avenir Next

ISBN 9781922332059 (PB); ISBN 9781922332066 (eBook)

A catalogue record for this book is available from the National Library of Australia

The Skywriters Project and the publisher gratefully acknowledge the financial support provided by Regional Arts New South Wales.

Contents

Acknowledgements

Front Cover Image: Niall MacNeil, "Celestial Emu rising over Wattle Flat"

Back Cover Image: Niall MacNeil, "Star trails around the South Celestial Pole from Wattle Flat"

Book Design: David P. Reiter

Photographer's notes on the front cover: The Milky Way rises to the east of the historic gold mining village of Wattle Flat, NSW, over a stand of Eucalypt and Wattle trees, bathing in the sky glow from Sydney 200 kilometres away. How many people realise that looking at the Milky Way like this is to look at a galaxy from the inside? Our galaxy. The dark dust lanes that form the aboriginal constellations of the emu and the kangaroo are superimposed on the combined light of hundreds of billions of stars. The bright bulge at the lower left just above the horizon is the very centre of the Milky Way galaxy. The brightest celestial object in the sky is the planet Jupiter. This image was produced from captures made on the 28th of April 2019 at approximately 11:15pm, using a Canon EOS 5D Mark II camera, and a Samyang 14mm f/2.8 lens on a stationary tripod. 30 x 20 second exposures were taken at ISO 3200 and the stars from each image were registered/aligned before being integrated to produce a high quality low noise image of the nightscape. Due to the apparent movement of the stars, the foreground was produced separately from unregistered exposures before being combined with the image of the sky.

Dozens of organisations and individuals have contributed to the Skywriters Project and to the production of this anthology. My thanks, first of all, to all the writers who submitted their skystories, and to my fellow curators, Suzie Gibson, Val Clark and Gai Lander, and our publisher, David Reiter and his team at IP. It was a great pleasure working with you all. Thank you.

My thanks, too, to our many collaborators and supporters without whose support this book might not have been possible, including

- All the Inland libraries that have hosted Skywriters events since 2017
- Arts OutWest Inc. (Bathurst)
- Outback Writers Centre (Dubbo)
- New England Writers Centre (Armidale)
- Orana Arts (Dubbo)
- Wiradjuri Study Centre (Condobolin)

- Red Room Poetry (Sydney)
- Research School of Astronomy and Astrophysics, Mt. Stromlo Observatory, Australian National University (Dr Brad Tucker)
- Donna Burton ('Donna the Astronomer'), Milroy Observatory, Coonabarabran
- Writing New South Wales (Sydney)
- Parkes Shire Council
- The Dish Café, Parkes Observatory
- Centre for Cultural and Creative Research, University of Canberra
- Lachlan Inn (Bathurst)

As well as the all-important in-kind support, we also received grants from

- Regional Arts New South Wales through our partner organisation, Arts OutWest Inc.
- Copyright Agency's Cultural Fund through the Wiradjuri Study Centre

Our thanks to all of you.

We acknowledge and respect the Traditional Custodians on whose Lands we live, learn and work.

Foreword

For this book, forty-nine Australian writers have gazed at our big Inland sky and imagined new narrative paths to connect Heaven and Earth, our planet with its Universe, and our inner worlds with the great beyond. Most live or have lived on isolated farms, in country villages and towns or in the small cities of south-eastern Australia's Inland and know this region well. Others have visited the Inland from their homes on the continent's coastal fringe to be inspired by our vast daytime vistas and the full cosmic glory of our night skies which can, of course, only be experienced now from the most unpolluted and sparsely populated places. I am very proud to introduce you to these writers' work.

Dark Sky Dreamings emerged from my Skywriters Project, part of the Big Skies Collaboration to catalyse new cultural and other opportunities for rural and remote communities in Inland New South Wales and the ACT. Over the past three years, I've travelled thousands of kilometres in Scarlet O'Barbara, the old red Toyota wagon gifted to me for this project by a literary friend. Scarlet is big enough for me to throw my camping gear in the back, and even sleep in her when conditions outside get rough. The Project has been conducted on such a miniscule budget that the amenities Scarlet provides have been fundamental to its success, as has the support provided by our many project partners, collaborators and friends.

I've met well over a hundred Inland writers at Skywriters events hosted by public libraries in Narrabri, Warren, Gilgandra, Coonabarabran, Dubbo, Parkes, Forbes, Condobolin, Grenfell, Cowra, Orange, and Bathurst, and at gigs hosted by Parkes Shire Council, Dubbo's Outback Writers Centre, Milroy Observatory and by local supporters in Molong and elsewhere. Skywriters newsletters and social media posts, including our call-out for submissions to this anthology, have reached hundreds more people within and beyond our region with the support of our partners, including New England Writers Centre, Writing NSW, Outback

Writers Centre, Red Room Poetry and the Wiradjuri Study Centre, and many individual supporters, including astronomers and astrophysicists.

The works in this book are as diverse as the authors themselves: funny, inspiring, thought provoking, poignant and profound. You'll find intensely personal memoirs, essays that evoke a deep sense of home and belonging, Sci-Fi fantasies that take you to freshly imagined worlds, and poems of transcendent beauty about our place in the Universe. Some skystories are light and entertaining comfort food, while others address obdurate social challenges: domestic violence, misogyny, sexism, racism, white supremacism, religious fundamentalism, mental illness, animal cruelty, environmental degradation, rural conservatism, the impacts of colonisation on First Nations peoples, and, of course, our species' future here and elsewhere in the Universe.

These issues polarise people in Inland communities, as they do in communities everywhere. Whatever the differences that segment and isolate us, however, we Inlanders all share the same big sky, all experience the same sense of awe and wonder when we gaze at the stars and planets, and all have our own skystories to tell, be they ancestral myths that encode deep sacred and secular knowledge; reports on the latest discoveries by astronomers and astrophysicists using the many Inland research observatories; astrological prognostications and horoscope readings; spiritual beliefs about sky deities and mystical events; bush yarns about encounters with UFOs; futuristic speculations and fantasies about alternate universes; space travel or life on other planets; or lamentations and curses about the absence of rain-bearing clouds to end our present drought. In these difficult times, stories that unite us rather than divide are more important than ever.

This Project has exceeded my expectations in so many ways. It has produced this anthology. It has brought together a dispersed network of creatives who are passionate about nurturing the arts and sciences in Inland communities, promoting cultural diversity, and enticing star-deprived Metropolitans from the coastal side of the Great Dividing Range to experience the celestial panoply of our very dark nights. The Project has allowed me to brush up my networking and editing skills; engage with writers whose work I have admired from afar; meet Inlanders who, like me, feel compelled to write; and, most importantly, to give aspiring authors support and encouragement to revise their drafts to publication standard and develop the confidence and resilience they'll need to keep going—because creative writing can be hard and lonely brain-work, no matter how experienced or well published you are.

But the Skywriters Project has also produced some very unexpected outcomes. These include the Inland Astro-Trail concept first mooted at a Skywriters event in Parkes in July 2017 and the still-nascent community organisation, Inland Astro-Trail Inc., which emerged from a Skywriters gathering in Molong later that year. The Inland Astro-Trail was conceived as an astro-tourism, cultural heritage, community development and STEAM (Science Technology, Engineering, Arts and Mathematics) outreach initiative to catalyse cultural, social, economic and educational opportunities in south-eastern Australia's rural and remote inland. Destination Network Country & Outback (DNCO) is now developing the astro-tourism component of this concept as part of the State Government's commitment to increase the number of visitors to the Inland. DNCO's consultants are now preparing the first Night Sky Experience Masterplan. We hope that many of these future astro-tourists will buy our anthology and be inspired by our stories to view both the Inland and the Universe in new ways—and even author their own skystories.

Another unexpected outcome of the Skywriters Project is the Condo SkyFest hosted by the Wiradjuri Study Centre in the remote little town of Condobolin. The SkyFest concept arose from the Wiradjuri Skywriters Project to encourage First Nations locals to record skystories in their own ways. It soon became apparent, however, that the impacts of the invasion and colonisation have been so extreme in and around Condo that few ancestral skystories have survived. Local Wiradjuri people including Tennille Dunn, Marion Packham, Bev Coe and the fibre artists of the Condo SistaShed have, nevertheless, found creative ways to revive, interpret and share ancestral skystories to inspire what could be called a cultural renaissance in their community. I feel very privileged to have been able to mentor these women and help them extend their support networks through our Big Skies Collaboration. You'll find Marion

Packham's memoir, "Riverbank Dreaming", in this anthology.

The Project has enriched the lives of its participants in so many ways. For me, it has been a delight to meet so many Inland writers and witness their cultural contributions to their communities, and to engage remotely with writers in other parts of Australia. And what a privilege it has been to work with local librarians and Council staff, with our many other in-kind supporters and partner organisations, with my fellow curators, and, most especially, with our publisher and Big Skies Collaborator, David Reiter, and his crew at Interactive Publications in Brisbane. My thanks to you all. The skystories in this book will resonate in Inland communities and elsewhere for many years to come. Who knows what other unexpected outcomes they will inspire?

– Merrill Findlay
Forbes, New South Wales
October, 2019

Michael Andersen

The Red Star

Away from the lights of the cities and towns, night happens quickly in the mid-year months, and June nights can be cold—but not this one. An unseasonal northerly breeze had brought an unfamiliar warmth for this time of year. Some of the locals believed that a northerly is often followed by rain, others that a downpour is more likely after three frosts. Still others, many of them First Peoples, reckon that you can estimate the number of days before the clouds drop their load by counting the stars in a halo around the moon. On this night, though, there was no threat of precipitation.

It was a weekend, and a little girl from the city was spending it at her grandparents' hobby farm. For her, the feature event was night time when she could spend as long as she liked under the celestial canopy. Her grandfather told her stories about the stars, the planets and the constellations. She loved them, especially his star stories. This child and the old man, her mother's father, were kindred spirits. She liked what he liked, even down to what her Nan put on their dinner plates. As Nan had commented many times, "You two are twins born sixty years apart!"

On this night, the 'twins' were sitting on a log that had once been a massive gumtree and were going through the ritual of naming the stars. With a cry of delight, the child pointed out the Southern Cross high in the winter sky. She could see the Pointers then scanned the firmament for The Saucepan. Her grandfather had explained to her that this asterism was merely the belt and sword of the constellation Orion, but she wasn't convinced.

"No Saucepan tonight," her grandfather told her. "It won't be in the sky until after dawn at this time of year."

Every so often, to the joy of both observers, a shooting star left a fleeting scar of light and, for a couple of hours after sunset, they could see satellites tracking across the sky.

"What is that bright star, Grandfather? Is that the Evening Star?"

Securing his balance on the log, the old man looked up.

"No, Darling. That's not the Evening Star. The Evening Star is which planet?"

The little girl inhaled deeply and held her breath while she tried to remember.

"Venus," she said.

"That's right. And, as it happens, Venus *is* the Morning Star. That one up there is Jupiter, the largest planet in our Solar System. A huge ball of gas about one thousand times the size of Earth. It has over sixty moons. Imagine that!"

The little girl shook her curls and gazed at her grandfather in amazement. She made herself more comfortable on the log and quietly hummed a familiar tune: "Twinkle, Twinkle, Little Star" …

Fran Bailey loved her work at the observatory. She especially loved the night shift when she could gaze through the big telescope that was so modern way back in the early 1970s. The mathematics of radio astronomy and the educational side of things also excited her. This passion for astronomy had been ignited by a visit to a planetarium as a schoolgirl.

Her colleague, Saxon Flynn, had been at the observatory longer than anyone could remember. He was affectionately called Gal by the scientists, technicians and even by the cleaning staff. He didn't mind the nickname. Not at all. He saw the diminutive of Galileo, the 'Father of Astronomy', as the biggest of compliments.

When they were rostered together at night, Fran and Gal would often play games to keep their eyes and minds active. Some nights their game was as simple as keeping count of meteors in the sky. On other nights, it might be something to do with moons. On one night, early in their shift, a technician blurted out a single word while he was calibrating a piece of equipment. "Red," the techie said. So red became their theme for that night's game.

"Grandfather, look, look, it's Mars, I know it's Mars—because it's red. I was learnt about it in school."

Her grandfather smiled and looked to where this seven-year-old astronomer was pointing. She reminded him so much of her mother at that age.

"Two things, little one. The first is you were taught it at school: taught, not learnt. You've learnt that Mars is red. The second is that it is *not* Mars! Mars is a planet; that is a star, Antares. It looks like Mars, though, or Ares, as some ancient people called it."

"Oh!" was the child's initial reaction. "Yes, it does look like Mars."

"Very smart of you to notice. Antares kind of means the "same as Mars"."

The moon was making its way towards the western horizon, a mere fingernail clipping. The stars were shining so fiercely they seemed almost close enough to touch.

"Grandfather, why are there stars?"

"Ah! They are more than just stars. The night sky is God's jewellery box, and each star is special, just like Antares, just like Mars."

"Tell me about Antares, please?"

Nocturnal insects were buzzing pizzicato around them. A mopoke called and, from across the darkness, its mate replied. The old man lifted his granddaughter onto his lap. His elderly collie flopped down at his feet to listen to the story, too.

"It was the time when the stars were young, a time long before people populated Earth," he said. "The sky would have been vastly different then from now. Comets and meteors would have been zooming through the voids between worlds, the moon was yet to be pockmarked by cosmic collisions, and all the constellations were still vying for their places in the cosmos. Orion, the great hunter, was yet to be named and the stars of the constellation we know as Scorpius dominated our part of the sky. The Scorpion was king of this section of the heavens—a harsh and cruel king. His very nature was mean. He was ready to attack the weak and defend himself against the strong. He hunted alone and took what he wanted whenever he wanted it. His wide claws seemed always ready to snatch his quarry, his tail always ready to inflict his sting. Not surprisingly, he had no friends amongst the stars.

"This was so long ago that many of the stars were not where they are today. The planets were new and still excitedly working into their elliptical routes around the sun. Some stars, like Antares, were restless. She roamed the galaxy freely as if in wonderment.

"Mars, although small, was young, strong, and brave. A soldier and a commander of soldiers, he guarded the rocky planets with the help of an army of asteroids, which stood in permanent readiness around the sun. From a great distance, he noticed another red body in the cosmos. Never had he seen anything as beautiful as this crimson star. On one orbit, he saw Antares paying him particular attention. He called out to her. She answered. Soon they became friends.

"Like all young friends, Mars and Antares enjoyed playing games together. Antares would try to be in a different place each time Mars completed his orbit around the yellow star he was bound to by gravity. As time and orbits went by the two cosmic bodies grew to love each other—the planet of solid rust red and the star that waited for him as he circled his sun. Mars would try to complete his orbits faster but they always took the same time. Antares always waited for him.

"Without them noticing, the stars and other planets were settling into new habits, too. On Jupiter, Mars' big neighbour, great storms were raging, while on his smaller neighbour, our own planet Earth, land and water were fighting each other for dominance. Dust and ice were forming stunning rings around Saturn, and lazy Uranus was falling on his back. Uranus remains that way today. Mars and Antares didn't notice these changes, though. All they saw was one another.

"But while Antares was waiting for Mars on one of his orbits, she noticed that the scorpion was stalking her as a cat stalks a sparrow. She fled across the void. Was there anywhere to hide? Where was Mars, her champion? Waves of darkness lifted her through space as the giant predator moved closer. Meanwhile, her suitor was on the other side of the sun and couldn't see her. Enormous pincers reached out towards her and missed. She rushed across the cosmos looking for shelter, but there was nowhere to hide. The odds were against this beautiful red star.

"The scorpion moved closer, and closer. He was toying with his prey—and then he pounced. He soon wished he hadn't. Making a meal of Antares was a miscalculation.

"Mars searched for Antares when he returned to her part of the sky but couldn't find her. Without her, there was more emptiness in the firmament than ever, more emptiness in his heart. He cried. Two tears fell and stayed with him. Aeons passed but Mars never stopped looking for Antares."

"But Grandfather, we can see Antares and we can sometimes see Mars. Why can't Mars see Antares?"

"Didn't you listen to the story, little love? Scorpius ate Antares. If you look just ahead of her, you can see the outstretched claws, and just behind his head, you can see Antares sitting in his throat. Keep following down and you'll make out the scorpion's stinging tail.

"You see, Antares became stuck in Scorpius's throat. The giant scorpion had never devoured anything as hot or fiery as Antares before. In her efforts to save herself, she had stoked up her inner fires. The burning was intense, way beyond anything Scorpius had ever suffered. Each night he twists and turns his way across the heavens trying to dislodge Antares. But he can't! You see, scorpions can't cough! Antares is therefore stuck inside him forever. And she is forever burning his throat.

"As for Mars, well, he spends his time wandering all over the night sky looking for his lost lover. He doesn't know she was eaten. That's why we'll always see Antares inside Scorpius and why Mars is never in the same place. He is still busy searching. And those tears he cried, they became Deimos and Phobos, the moons of Mars."

The two astronomers were in deep discussion about the red planet. The Rovers were sending back data that suggested the presence of frozen water on Mars' poles. Was it possible that this dead red planet could support life? Facts are what astronomers deal in, though. Gal and Fran knew it was not their job to speculate about extra-terrestrial life; nor was it their job to search for it. The possibility was nevertheless intensely fascinating.

While they were speculating about life beyond Earth, Fran was observing Scorpius and the red giant beating like a heart in the constellation. Despite her education and understanding, she could name only two or three constellations that resembled their namesakes. Scorpius was one of them.

"Tell me what you know about Antares," Saxon asked. "Come on, Fran, surprise 'Ol Gal."

"For a start, there's nothing I know that somebody somewhere else doesn't already know," Fran replied. "Antares has been admired and pondered for millennia, and there's still conjecture about her name. The popular theory, as you know, is that it means 'similar to or rival of Mars'. That's from the Greeks. I like the possibility of it being named for the ancient Arab hero and poet Antarah ibn Shaddad, though. Of all Arab warriors, he was considered the bravest. Many songs and poems have been written about Antares, of course, some dating back hundreds, if not thousands, of years. All civilisations and tribal peoples had names and legends about this red beauty.

"The star itself is huge with a radius not far short of nine hundred of our suns. Because it's a red giant, it's likely to go supernova. Should that happen, if it hasn't already, it would be over six hundred years before anyone on Earth would know about it. As far as brightness goes, Antares is about the seventeenth brightest star in our sky. And she is not alone. Antares has a companion star called Antares B. It was discovered in the early 1800s during an occultation of Antares by our moon, but it's difficult to see under normal conditions."

"I've heard some yarns in my time old man, but you've outdone yourself this time with that one!"

Both the storyteller and his listener were startled. They'd been so involved in the story that they hadn't heard Nan walking up behind them.

"Nan, Nan I feel so sad for Mars and for Antares. That Scorpius is horrid. I hate him for what he did. Oh! It is such a sad story."

"Yes, sweetheart, it is, but it's just that, a story."

Nan handed her husband and granddaughter a cup of cocoa each. "Ask your Grandfather for some truth about your stars," she said, with a smile.

"Please, Grandfather, please!"

As they drank their warm nightcap, the lesson began.

The old man explained that the universe is huge, that our Solar System is just a speck in the mass of stars we call the Milky Way Galaxy. Scientists estimate there are billions of galaxies in the universe, and each galaxy is made up of billions of stars, planets, asteroids, comets and things we've not yet discovered.

"Mars is much smaller than Earth and, like Earth, it's a planet. Antares, on the other hand, is a star. Just like our Sun is a star."

"No, the sun is the Sun," the little girl insisted. "It isn't a real star, is it?"

"If you could stand on Jupiter and gaze at the sun, it would look like a star!" her grandfather said.

What really amazed her was Antares' size. "So, if Antares was where the sun is, we wouldn't be here, and Mars wouldn't be here also?"

"That's right, little one. In the overall scheme of things, the sun is quite small. The moon, of course, is much, much smaller again. Due to its position in the sky and its distance from Earth, it appears to be the same size as the sun. That's why we have solar eclipses when the moon comes directly in line between the Earth and sun. When that happens, day turns into night. Ancient folk were very superstitious about eclipses. They didn't understand so they feared them."

The child put her empty cup on the log and climbed from her grandfather's lap to her grandmother's. Snuggling into Nan's warm softness and cradling her head on her grandmother's bosom, she urged her grandfather to tell her more about Antares. But her eyes lids were losing the battle to stay open.

"Well! Surprise Old Gal you did, Frances, my girl. You really know your subject!"

Fran smiled at Gal and modestly observed that there was much more to learn.

"Would you like to tell me what you know about Mars, then?" she teased.

"Oh no, dear girl, I respect you too much to burden you with an old stargazer's ramblings."

Fran took the subtle hint that the game was over and that they should settle into the work they were being paid to do.

"I'm impressed with your understanding of Antares and the way you imparted what you knew," Gal said later. "If you can't explain something to somebody then you really don't understand the subject yourself. You'll go far as an astronomer, Fran, but you could do big things as a teacher or even as a professor too. I believe that. Give it some thought!"

A nudge in the ribs from Nan reminded him it was bedtime for their granddaughter who, by now had succumbed to slumber. He stood and carried the child to bed. Nan collected the cups and followed.

"She hangs on your every utterance, you know," she said. "Our little girl is very bright, no doubt about that, and keen to learn. Mind you all this star and space stuff is still rather beyond her, but she tries."

"The pleasing thing is she's willing to listen and learn," Gal said. "She knows the stories aren't true, but she learns from the facts she picks out of them."

"Saxon, why don't you take her to the observatory one night? She'd enjoy that. Fran could spend a little time with her and explain the sky in simple school-girl terms."

"I just might do that, Nan, I just might do that."

While she slept, the little girl dreamed. She dreamed of Mars and his two tears of heartbroken sorrow, but most of all she dreamed of Antares. She desperately tried to warn her about the giant scorpion hunting her in the sky. She didn't want that nasty King of the Sky to catch and eat her beautiful red star.

Michael Andersen lives under the wide-open skies of Northern Inland NSW, and, in his own words, "has spent quite a few decades orbiting Old Sol". A career in country broadcasting has allowed him to work in many towns across Australia. He suspects that the hours he has spent conceiving and writing 'creative' radio advertising 'ignited a spark' to write his own stories for his own enjoyment. He wants you to enjoy reading his words as much as he enjoyed writing them. He lives in Moree, New South Wales.

Damian Balassone

The Galaxy for Our Eyes

Behold celestial skies,
the galaxy for our eyes,
my love lights up the darkness like a luminescent kite.
Behold celestial skies,
the galaxy for our eyes,
her thrilling touch miraculously delivers me from night,
and all who see her laughing eyes are bathed in beams of light.

Two Mirrors

The love that cuts through lies
shines down from lunar skies
and mirrors in the streams
the image of a dream.

The image of a dream
emerges from the streams
and mirrors in the skies
the love that cuts through lies.

Solar Eclipse

The moon has parked her shiny arse
between the earth and sun,
she's obviously oblivious
to all that she has done.

Grace

When darkness descends from the sky
and Grace is pursued by a lie
she summons the stars and the spheres
and glimpses the moon through her tears.

The Night Gardener

As Rocco waters his geranium
the moonlight dances on his cranium.

The Scientist

Synagogue, temple, mosque, church
did not at all reward my search,
so I renounced religion and became free
and now I study the stars and the sea.

Damian Balassone's poems have appeared in more than 100 publications, most notably in *The New York Times*. He is the author of three volumes of poetry: *Prince of the Apple Towns, Daniel Yammacoona* (Ginninderra Press) and *Strange Game in a Strange Land* (Wilkinson Publishing) with another collection The Book of Original Clichés, forthcoming.

Tatiana Bonch-Osmolovskaya

arc

night cries with splinters of feathers
of a celestial diamond emu
extending on high legs to the top of the sky
lengths from horizon to zenith to horizon
widths to embrace us all
until the dawn comes

celestial date

on the eve of eclipses, and comets' fall,
and parades of planets, and other
harmless but awe inspiring phenomena,
people take out their sooty glasses,
binoculars, cameras, telescopes,
stand still on solid ground in safety;
with optics to shorten the distance
people go outside to meet a star,
to see the fading sun, or a crazy comet.
as much as I remember the circumstances –
eclipses of the Moon and the Sun,
passing of the shadow of Venus,
Halley's Comet, and large blue Moon,
and other celestial phenomena
always occur in rainy and cloudy weather.
the skies do not show up for a blind date,
a narrow hand closes the carriage curtain,
a wanderer pulls the hood down,
you stand holding in hand
a sooty glass and camera,
alone on top of the world.

Tatiana Bonch-Osmolovskaya studied physics at Moscow Institute of Physics and Technology and philology at Moscow State Humanitarian University. She is an author of thirteen books in Russian, including *Introduction to the Literature of Formal Restrictions* and a novel *Through layered Glass*, and co-editor of the anthology *Freedom of Restriction*. Her poetry in English has appeared in *Can I tell you a secret? Across the Russian Wor(l)d, Bridges, London Grip, POEM, Rochford Street Review, Journal of Humanistic Mathematics,* and other journals. She is a member of Board of PEN Moscow and editorial committee of Russian literary journals: *Another Hemisphere* and *Articulations*. Tatiana now lives in Sydney.

Tom Bristow

anthemic australian country constellation

if you really want to know, it was just after the third week of the equinox when he said to his mum:

> I'm thinking of using it as a device, you know, a conceit,
> for a play or a novel or something … on reconciliation.
> going with the idea that conciliation comes first. do you
> know what I mean?

she knew what he meant. they had recently completed a unit in European-Australian legacies for an online BA at James Cook University, but they were eager to start all the new creative writing minor units. they completed assignments together, but were enrolled for all formal processes in his mum's name because Lachlan was too young.

> let's drop the idea we had about setting our next piece
> along the historical pastoral route in the town, remember,
> we discovered that it's named after the Secretary of State
> for the Colonies, part of the story about getting through
> the mountains … let's avoid that, but, well, I'm not sure
> of the next step … we could put some of these thoughts
> in a document that looks like a prose poem, as if it's a
> treatment of sorts for a larger piece, a micro-novella
> maybe. which it might be, later. thing is, mum, I'm just not
> sure if it's best to wait for these ideas on form to develop,
> or for the writing to grow in my head in its own way. how
> do I know what's good. And … if I start writing, mum,
> where will it take me?

her composure settled (with the demands of the voicemail inbox on her phone despatched some minutes ago):

what is it … about, Lachlan?

the reply was immediate:

> y'know the Southern Cross has been rotated about 90
> degrees clockwise—from around 8 o'clock to 12—on the
> Australian flag?; yeah, well, I like to think of the blue as
> the sky at night before it moves into darkness. and, if you
> imagine the stars as they really are, in the sky, when you
> are lying down to sleep tonight … ok. mum? close your
> eyes when you are ready and gently put your left arm on
> your heart, relax and let your right arm fall freely to your
> side slowly, as if it were gesturing outwards, like this.

he moved his body as he described and continued:

> then you've got the same shape in your body as that of the
> elongated cross on the flag; you are mirroring each other,
> if you can imagine! it's great! your left elbow is the tip of
> the star or the bottom part of the shortest arm of the cross
> as we see it up above, and the hand on your heart is the
> smallest or farthest star … got it!?

the features of his young face opened into one large red smirk as his
mother took immediately to the carpet and synchronised her arms
movements with his.

> so *that's* the idea. finding a way to magnetise yourself to
> the stars. I've been experimenting, and this is a bit weird
> so I don't know how we'd put it in words, or get away
> with it in words: if you kinda meditate, you … well, you
> might experience what happened to me … last night my
> mind moved through the sky as we've just tested, and I
> felt something special, a new feeling, like honour as I was
> freeing myself from the earth. yes! I was able to kinda
> watch it develop, with my mind kinda thinking about it *in
> words* whispering to myself, you know with the voice in your
> head. my mind was almost empty and had space to process
> things like this, so I reckon it kinda decided to think of this
> emotion, or develop it into focus, this honour, on its own;
> why it chose to add words to it, I don't know. but in adding

these silent words to what was happening, in my head, you know, not out loud … kinda to myself to make sense of what was happening, it forced me to put these words and the feeling together. it was nice to narrate the feeling as it emerges rather than name the feeling. and that in itself is what you'd try and register in the name for the feeling: honour is close, an honour with permission given by the stars for the retelling and passing on of the feeling. I'm losing my reasoning now, eh? kinda rambling, and not sure if art can bring it all together …

go on, darling

well it made me feel that we can connect to something larger that we are all a part of, y'know, so this honour or pride in that empty or unnamed parts of the feeling lead us to a sense of … responsibility? … for this connection, yes, maybe that's it. it's still coming to me now you see. And then that's enough generalising. we'd have to write the next bit separately then join it together. don't tell me what it does for you until you've written it down and until I've written my bit down. Ok? for me, the feeling connects me to the river and what happened out there on the land, to responsibility not only to keep our connection to the stars, which is massive, isn't it!? the massacres, poor land use and now coal and stuff. but the bond or connection, is much mightier than all this short history of mistakes, to which we will come together to dismiss. like the light, this new feeling comes to us after it has travelled over all the other people and creatures living under this constellation throughout all history. how? because they are so old those dying lights out there, which made me think the cross is a way for us to start thinking about caring for all things if we want it to. mum?

he wanted to share how strange this feeling was at first, and then go on to share how it began to make sense: "for me, the feeling I experienced during this process led me to think about forgiveness"—but he didn't as he wanted to keep that to himself right now. forgiving trespasses. you know what I mean.

*

That story is a first attempt to show how things really are things, like stars are fat and heavy and dying and out there. They are, aren't they? Take a look. We might infer from this short performance by our courageous schoolboy speaker that Lachlan likes The Saucepan in the way his father loved The Plough. I can't say; I only know he loves looking through Orion to the centre of where we are. Good night.

Aussie poem #29

It goes 'zzing gg' when you switch it on
'zzIing g' kinda thing: the perfect blue desk lamp that is now on
my desk bringing light as I write
I picked it up yesterday morning for only 5 bucks in the 'Op-Shop'
right on the bank of the Macquarie River;

so many things that are still here
in you
that sparkle;

Australia 'zz**II**ng'
like the 1980s blue and zing of the skies and the lamp,
I share with you;
'zzing' the stars
as I lean out of the Brumby
as I drive to Bathurst from Perthville under raining stars

Zing, it's the go! so
goodbye Perthville, New South Wales
see you again for you are
the light at the end of the tunnel.

It is to you, Australia!
the Australia that I know from the regions
shaped by folk
        – good folk that I
        love to see when I go out and about –

to you: I offer my deepest gratitude
to you my lifelong respect,
my commitment,
            and good intentions.

7 March 2019

Here comes the rain, stars

And you've done nothing about it

I remember a poem by an American last century
About the Sun at Fire Island
Not at all cosmic, but a nice relationship between an earthling
and white dwarf star (read like it was real
Northern-hemisphere stuff, y'know), we should imitate this.

You know, stars, I really think it'd be good for your profile
If you, well, y'know, did a little more about the weather
Or 'climate', yes, let's use that term …

Stars! Wake up. What is it with you!?
You are out there
        dying while we're trying to live under you.
Seriously, I was saying that it's dry here. OK? And we'd
Like some rain. Understand? Can you help? And, well,
I have to say it:
If we can't pray to you then
We'd better do something else.

But I guess we could learn from you. If there's enough time.
(Y'know, between us, I don't buy the 'moon-planting stuff' but
that's 'cos I haven't tried it …) and we can't leave it to historians
and agriculturalists to tell us everything, eh? What else can you
give us other than your death to understand that the planet is
alive?

In response you won't say anything.
Seems unfair. Especially when we're trying to grow the
Country. Y'know, 'grow' like a meta …

– quick, inside! It's coming down. Now!

The stars in one tiny speck of rural NSW tonight

… are huge. As you'd expect in country.
Last week they told me to "sort out your finances this month"
So I'm doing it. Thanks, stars!

They have guided us and many others like us looking for home
     – like trade winds gone viral –

Thankfully they don't use social media, the stars in the Central
Plains; no,
the stars in the Central Plains just the right side of
     Blue Mountains do this:
        they communicate through the crickets
     and the silence between race meetings under sacred
        misnamed mountains

You bet they wheel past sometimes as I'm wheeling under them
We're calibrated, the stars and I, to each other

… although, to keep it honest: I don't really know the
     angles and speed of light involved in what I'm
     saying
I don't want to know all that although it'd be good to know it

I simply hear their tears
     of joy
        and in this joy
           (I know, don't worry, I know
              we have problems under these stars

               – Australia is an exception
                  in some contexts
                      and not so in others)

we leave the past to one side or we won't see the stars for clouds.
Perhaps I'm wrong, but if we think about the sun's energy in the

plankton that becomes oil, or the ancient energy stored in those old forests that somehow return to us as coal, then we're always in an old world, there's nothing new-fangled about 'electricity', just like there's no new water. You follow me?

We miss all this if we put clouds out there … Although – crying for rain as we are, we want more clouds. Let's seed clouds, stars! And then, you're with me, stars, keep up – some clouds are actually Cosmic dust clouds, aren't they? But you gotta see them at the right time to know this. Some of these new stars out there, so we're told. A lot of it begins with dust. So-called dust.

So, stars, sorry; you are godforsaken tonight. Although I don't know what that means, either.

Tom Bristow is the Roderick Research Fellow at James Cook University, Queensland, an Honorary Research Fellow through the Australian Research Council Centre for the History of Emotions at The University of Western Australia, and Research Fellow at the Institute of English Studies, University of London. He is also editor-in-chief of the journal *PAN* (Philosophy Activism Nature), and the Literature, Culture, Media environmental humanities series editor at Routledge. Tom composed these poems in Bathurst, New South Wales, on a Marie Curie Fellowship with Durham University. He now lives in North Queensland.

Val Clark

The Demise of Unity

A Chronicle of N'arth myth

The searing emptiness of loss stabs into my gut. This is not my world. The meagre stars beading the midnight sky above me, above N'arth—New Earth—they are not my stars. My stars, on the Western Plains of New South Wales, charge across the sky, from horizon to horizon. Out of habit, I raise a hand to these alien stars, ready to thread my fingers through their brightness.

Out there somewhere, sometime, is my home, Earth. But it's a paradox, time and space running forward and out.

I drop my hand to my chest and spread my fingers, pressing down hard but nothing stops the pain of permanent separation from my parents, my friends, my world. Instead the wound grows and I imagine myself leaking out, soaking into the foreign ground.

My heart jolts and the breath in my chest judders. A little death.

My hands reach out beyond the swag, scrabbling for round lumps of rock. Surely here, in this place of bare rocks and boulders, there must be healthy, healing *HeartStones* and the *Prime*. How had I forgotten the *Prime*, the voice of sentient N'arth?

Healthy, healing *HeartStones*. A trail of wellness soaks out of them, filling the broken spaces.

Welcome.

My fingers clench the *HeartStones*, frantic not to lose this connection with *Prime's* voice.

I lift one to my face. The night is so dark now it's invisible but, as I roll one over my soft cheek, the tension dissolves. My shoulders relax. I tuck the stones into my belt pouch. I think, for the first time since Rad left, I just might be able to shrug off this hollow weariness and sleep through the night.

Rad, the *Hayyo'im* runt of a race so damaged their bodies flickered in and out of this dimension. Rad and me. An unlikely pairing, sharing our minds, our hearts, our souls. Or at least we used to.

I'm about to roll over onto my side when Brghyn drops his swag, stretches it out and settles down so we are head to head. 'As a boy I would come here, out into this wilderness, to be one with the *Prime* and the stars.'

I point to the constellation directly above, six stars in an almost perfect circle. 'Tell me about that one.'

There's a long silence and I wonder if I've offended—crossed a cultural boundary Brghyn's race holds dear. He surprises me by shifting so close our heads bump and the beads in our plaited hair touch and gently click. I stay as still as I can, savouring this rare intimacy.

'We are not a prosaic people, Shannon. We don't tell stories around the fire as your people do.'

Your people. No, they would never be my people again.

'But you do have some sort of story?'

'We call it the Ring of Arrival. The ships that came with your people thousands of years ago came through the Ring. Your *Hayyo'im*, Rad, has she shared this history with you?'

I roll over, hoist myself up on my elbows and meet his eyes—glowing like honey in the dark.

This is so hard. Rad has gone on a quest. Where once we were mind-melded, now there is the slimmest thread connecting us, and she has made me promise to leave her be. So, I have to search my mind, go back to the moment when her dying mother breathed her race's memories into me—memories I would pass onto her children, onto Rad.

There it is. I sit back and retell it in Rad's mother's voice as if she's putting the words into my mouth, which, in a way, she is.

'Murroomay are one. Mother and Father. Life givers to this world, the world we now call N'arth. But, in those days, it was called Saranooshal. Murroomay crisscrossed the universe in a great orgy of creating lasting aeons, but, for them, mere seconds. One day Murroomay returned to sit on the circle of those stars and looked down again on Saranooshal.

Saranooshal was special. Unlike all the other planets, Saranooshal was alive. Sentient. Murroomay's first creation left to evolve and develop how it wished. When they returned, they found succulent rainforests, deserts, grass plains, seas, fantastic rock formations, animals, birds, insects, but no sentient beings.

To the unspoken questions, Saranooshal replied, 'We did not people this world, Murroomayo. We waited for your return— waited for you to create for us creatures who will care for this land.'

Murroomayo sat for many more aeons on the Circle of Stars pondering what this creature would be like. Not just its appearance, but its characteristics. Finally, they called this new creation *Nurture*, populating the world with them. With tough, wiry *Nurtures* in the wild desert places to fantastic fish-like *Nurtures* who dwelt on land and in the sea, Murroomayo.'

Brghyn breaks eye contact sits up and turns to face me. 'It is a myth, a lie. There are no such beings as *Nurtures* on N'arth. And, how does a world speak?'

'You and me, we've heard the voice of *Prime*.'

He doesn't argue.

I look down, playing with a stray thread on my swag so that he can't see the tears pouring down my face. The rest of the story is here in my head, the story about the destruction of the *Nurtures*, but I can't say it in Rad's mother's cold and dispassionate voice. I translate them into something warmer.

'The *Nurtures* gave sanctuary to my race, the *Hayyo'im*, the most advanced race in the known universe—four-legged with an extra pair of hands and an unparalleled ability to control and manipulate the physical world. They built the city of Pleth, the Temple of Light, the canals and many more wonders. In their search for knowledge they—'

A sigh racks Brghyn's body. 'What happened to the *Nurtures*?'

I'm glad he believes me, experiences their loss as as I do.

I lay back on my swag. I don't remember getting the *HeartStones* out of the pouch but I'm grasping them against my chest again and reaching out to the *Prime*. Please tell me they are all right.

I pass a stone to Brghyn and we wait patiently until the *Prime* responds.

Murroomayo loved Saranooshal and the Nurtures *above all creation. Many times they returned to sit on the Circle and chat with Saranooshal about the good work of the* Nurtures, *overjoyed that Saranooshal and the* Nurtures *had welcomed the* Hayyo'im. *But their happiness dimmed when they realised the direction the* Hayyo'im *experiments were taking. Murroomayo had set the worlds in motion and could not interfere.*

Murroomayo *sat on the Circle until the dark night when a* Hayyo'im *device was activated, changing them forever and killing every last* Nurture. *Yet,* Murroomayo *was ready, drawing each* Nurture *life force up into the sky, through the Circle, and then showering them back.*

They are the stars dancing and singing in the sky above you. Forever dancing and singing.

Val Clark is an award-winning storyteller, visual artist, literary fiction judge and experienced creative and memoir writing workshop presenter based in Dubbo. Her novel, *Lost*, a young adult/crossover fantasy, is the first in her *Chronicles of N'arth* series. She is presently editing the sequel. A self-confessed writing workshop junkie, she started the annual writers' festival, WestWordsFest, in inland NSW five years ago to feed her addiction. Next to writing, trying to learn the penny whistle and travelling bring her alive.

David Clarkson

The Ditch

In my twenties, I made the journey across "the ditch" to live in Australia, as many Kiwis do. Australia seemed completely foreign to me at first. The roads were too big, the distances too vast, the temperatures too hot, and the skies too loud with the bright lights of what, to me, were vast megalopolises! It was on my first journeys to inland New South Wales and rural Queensland that I started to experience the beauty of Australia: the bush sounds as evening draws in, the clean air tinged with the scent of distant deserts, and, yes, those same dark night skies I knew from my childhood. The inland, on cloudless eves, lets you fall upwards into never-ending skies that are, literally, wonder-full.

My earliest memories are from rural New Zealand. From Fairview near the medium-sized country town of Timaru on the east coast of the South Island. We lived in a reasonably run-down old farmhouse with a coal range for cooking and an outside drop toilet that needed emptying twice a week. My father had run a prosperous business in Wellington, New Zealand's capital and second largest city. He was one of the best plasterers in town, or so the story goes, but his love of gambling and drinking sent the business bust. Hence the move to the outskirts of Timaru in less than perfect circumstances to a less than perfect house.

But I loved growing up in the country. At Fairview we had numerous fields to play in, vistas of both the ocean and the southern alps, farm animals, native birds and countless trees to climb. The nights were the best though. Vast pitch-black skies that positively glowed with stars and planets and views that took us to the heart of the Milky Way. Mum would often turn off the lights and take us out to the side lawn to spend what felt like hours laying on our backs and gazing into infinity. Sure, we knew a few of the constellation's names, and some of the planets—but the main game was just taking the time to be in total awe of the vastness of the universe. A joy without a name.

Most summer holidays we went to the South Canterbury High Country, the Mackenzie Basin, and stayed in a small bach[1] on the shores of little-known Lake Alexandrina, a spring-fed lake near the larger Lake Tekapo. The lake gave us great fishing and swimming— and, even better, great astronomical viewing. The night skies were stupendous there. We'd lay on the tussock and counted the shooting stars. It seemed that time was suspended in that heaven-scape.

Indeed, such is the wonder of those skies, that the entire Mackenzie Basin has recently been declared an International Dark Sky Reserve and a new Astro-Museum has been built at Lake Tekapo. I can almost hear my mother saying "I told you so!"

Mum passed her love of the night skies on to all her four children and seemed to have had the knack of being at the right place at the right time to see the most special of celestial phenomena. Late one night, for example, she was outside hanging washing on the clothesline when suddenly the whole side of the house lit up. She thought a car must have been driving up behind her—but that made no sense because there was no road. She turned around and there in the sky was a huge light hurtling towards her. It took up a quarter of the sky, she said. Having studied astrophysics at Canterbury University, I now know that she bore witness to a small asteroid passing extremely close to the Earth without being captured by Earth's gravity well. Mum watched it loom terrifyingly large and fiery in the sky and then recede at what must have been tens of thousands of kilometres per hour. "Out and out and out and up and up it went," she told us later. "I never knew space was so big."

Several years later, on one of my first forays into astrophotography during my first year at high school, I was up late photographing the stars out the front of our house, while Mum was inside hosting a party. The party ended, and she came out to watch the skies with me. And there on the horizon we saw three spherical objects moving at speed. They were orbiting one another and took about twenty seconds to cross our field of vision. They were fast, precise,

[1] In New Zealand, a bach (pronounced batch) is a holiday house.

relatively low to the ground and mechanical in a strangely organic way.

My mother and I had seen a huge array of night phenomena: evening stars, planes, geese, satellites, auroras, distant cars, eclipses of both sun and moon, meteorites and more. But what we saw on the horizon that night was none of these. We were far from any flight path and it was 1 a.m. in rural New Zealand, so what could it have been?

We were both convinced we had seen our first UFO and felt very excited. We knew, however, that no one would ever believe us! It was our secret. Regardless of whether we'd seen a UFO or not, we both felt that our frequent pilgrimages to the skies at night were vital to us. Our stargazing gave us a window into our own relative insignificance and into a universe that allowed us to bear witness to a small fraction of its unfathomable glory.

When I arrived in Australia many years later, it was as if the night skies had disappeared. The bright lights of Brisbane and Sydney held little of the night glamour of the wondrous Canterbury dark skies. That changed when I visited my friends Sarah and Mathias in the countryside out of Bathurst in rural New South Wales, though. One of the first constellations my mother had shown me was Scorpio. Its beautiful curving body and size make it relatively easy to identify. So, there I was, one cold autumn night near Bathurst—and rising before me was the mighty Scorpion. It was like a homecoming! Australia was no longer such a foreign country to me because there, in the sky above, were the stars of my childhood. Even though I was thousands of kilometres from my hometown I could now feel safe under the same night sky roof I had grown up with and loved. And those stars will continue to rise long after I am gone like sentinels to my mortality.

My mother is dead now, and, in a blink of time, I will follow her. Last week, on a visit to the countryside near Orange, I showed my daughters those same stars. My girls were in awe. So, thanks Mum for the night wisdom you passed on to me.

David Clarkson is a theatre director, producer, mentor, arts leader and innovator with a strong commitment to regional, rural and remote communities. He co-founded the physical theatre company, Stalker, which has enjoyed national and international success. His poetic, image-based, conceptual, performance work has been at the high end of physical theatre practice and has influenced many artists working in this field. David's work includes solo performances, professional ensemble work, installations, community outreach projects and Olympic opening ceremonies. His performance works have toured to over thirty countries and have been seen by hundreds of thousands of people. www. boxofbirds.net

Rosemary Curry

Brunhilde's Solution

Brunhilde smiled at the kids gazing admiringly up at her. She wasn't fond of talking to school children, but it was part of her contract. Let people see astronauts as ordinary people who get back to Earth safely; that was the instruction. And *don't* discuss Christa, her friend who died when the Space Shuttle Challenger broke up 73 seconds after launching! Nasty little school children had revelled in the TV replays of this disaster. "Come and watch Christa blow up again!" they shouted to their friends.

Hilde even disliked being in classrooms. They reminded her too much of her own unhappy school days back home in inland New South Wales. As a foreigner among children of mostly British descent, she had experienced daily name-calling: Fascist, Nazi, Wog, Dog's-body. 'Look what the cat dragged in!' 'Why don't you speak properly?' And she was repeatedly being 'Sent to Coventry', a phrase she soon learned meant ostracism.

"You always feel you're not good enough," she would tell her mother through tears at the end of each day. "I wish I were dead. Why did we have to come to this country? I want to change my name."

Her mother, Ilsa, was also excluded from the town's society because of her foreign accent and foreign-sounding name. She knew what it was like to be a displaced person who didn't fit in.

"Come on," she said. "We'll take some time off school and go to the State Library. It's time you learned a few facts of your own history. Be proud of who you are. Many from our dear Fatherland helped make Australia. We'll look them up. Your great-great Uncle Ludwig was one of them."

The Librarian brought them some of Leichhardt's papers and made them put on white gloves before they touched the precious relics. Hilda liked the smell of the old documents, particularly his maps. She could feel his presence as she ran her fingers over his drawings in the margins. Ilsa translated his German to inspire her daughter. "I want to vanquish the wilderness of Australia for the benefit of all mankind. I want to explore the New World," he had written.

They spent hours going through the explorer's journals. He loved "nothing better than camping out under the stars studying celestial navigation," they read. He had steered by these same stars until the day he disappeared. "His final journal must be mouldering in the dust somewhere," someone had noted in a margin.

"Has Providence deserted me?" Leichhardt asked in one of his final journal entries, as he faced more and more problems on the track. "I am fearful of losing God's blessing."

"But he kept on exploring and collecting specimens of everything he thought was new to science—and never gave up."

"My great uncle was a prince of explorers," Hilde whispered to her mother. "I'll be an explorer too."

"You can be," her mother replied. "Leichhardt was a migrant like you. He was mocked and taunted just as you've been, but he's someone to be proud of, despite what those who derided him said. You won't lose God's blessing, though. Just keep on trying to be the best person you can be. Persevere as he did."

Hilde went back to school and worked on improving her frame of mind. She learned to ignore her less than pleasant schoolmates and decided she would become a space explorer. When they shouted "Why don't you go down to the morgue and tell the man you're ready," she yelled back at them: "I'm going further than any of you. Just wait and see!"

She felt a weight had been lifted from her shoulders and was finally able to concentrate on her studies in preparation for her future in space. She breezed through her Mechanical Engineering degree and moved to the USA, where she successfully applied for a job in the aviation industry. From there she progressed to NASA, her final goal.

Like her great-great Uncle Ludwig, Brunhilde also kept a journal. After her first space mission, she described how she had gazed at the stars, watched meteors on kamikaze burn-out flights, and then gazed back at Earth, a tiny, fragile ball of life hanging in the void. "The most amazing experience of my life!"

"I could go on like this forever," she told her journal many months later, "but Hans, my fellow astronaut, is ruining everything. He's making every effort to destroy my will to live. He's insulting me, laughing at my accent. If I do something wrong, he explodes in an unprovoked rage and then won't talk to me for days. He tells me 'you don't speak properly.' It's just like being back in school. I've lost my way. I can't take this anymore."

Weeks later, another desperate journal entry. "I've decided. I'll put on my incontinence pants and the rest of my astronaut's suite, and shoot like a rocket along the highway until I hit something, just like the meteors I saw in space."

But then she reconsidered. "No, that's not good enough. I'll take him with me!"

Her diary entries became more and more desperate, more and more unhinged.

"I will end his life, too. He won't destroy me; I'll destroy him.

"Here's the plan. Next launch, 73 seconds after takeoff, I'll shoot off amongst the stars as Christa did. I'll explore the Universe in the next dimension. No one can thwart me now. Maybe I'll meet my mentor, Uncle Ludwig. Together we'll be unstoppable.

"I know how to do it. I feel relieved now that I've worked it out. I'll soon be free. Then I'll find out what's out there. I just have to wait for the countdown.

"100, 99, 98, 97 …"

I'm counting.

Rosemary Curry is a South Australian presently living in Central West NSW. She is as a Registered Nurse who, between having five children and running Family Education groups, studied Psychology at University. She is also an enthusiastic amateur palaeontologist, and, when not studying rocks and fossils, collects oral histories and occasionally writes articles for history magazines. In 1989 she published two small books based on her oral histories interviews. The most popular of these, *If That Man Comes Here I'll Shoot Him*, is about Australia's High Country Women. Rosemary now runs a creative writing group for U3A in Orange.

Garry Dean

The Real Deal

It started as a typical day. I got up, made coffee out of something that looked and smelt like the real thing, switched on the TV and sat down to watch the morning show. I had hundreds of channels to choose from, and there was something comforting in their fake cheeriness. If nothing else, it filled the emptiness of the house with the sound of human voices. The house was perfect in every detail, down to the cracked tile in the bathroom, but this wasn't my house. The real one was a long way from here. Sipping at my coffee, I began to ponder the days, weeks and months I had been here.

"You can't do this!" I had raged at my invisible captors, after being plucked from my real home, my world, my life.

"Stay calm, do not worry," came a voice from the ceiling. "You will be well treated."

"Bullshit," I replied. "You can't just take people and lock them up like, like animals."

They had seemed genuinely surprised at this. Hadn't humans treated less intelligent creatures in the same way?

I should have been terrified. I'd been abducted by aliens, and all I could do was get outraged. Anger had always been my response to fear. I was here because I was a typical healthy human female. My keepers knew all they needed to know about Earth and its dominant life form, but there was nothing like having an example of the real deal. I was the equivalent of a fish in a bowl, a bird in a cage.

"No bloody way," I growled. Going to the front door, I yanked it open. All there was beyond the welcome mat was a vague white nothingness. I slammed the door shut.

"Take me back!"

"Please, stay calm, it's alright," they replied.

I felt that familiar tightness in my chest, and a strong desire to break something.

"Why not sit down and relax?"

That silky comforting voice enraged me further and, as a typical form of lower life, I lost my shit! I grabbed a stack of plates from my fake kitchen and sent them sailing towards the windows. I didn't know if they were made of glass or not, but they shattered in a very satisfying way.

"Please don't excite yourself."

I overturned the dining table and pulled pictures from the walls. I swore, screamed and shook my fists at that disembodied voice until it fell silent.

At some point, I collapsed in a heap amongst the chaos of the lounge room and wept. When I woke up the next morning, I found everything back in its place, as though nothing had happened. Over the following days, I grew hoarse trying to reason my way out, but they politely refused to release me. I was stuck here— wherever 'here' was. There was always food in the fridge and water in the tap, and every day, the light through the windows brightened towards daylight and dimmed towards night. I felt no threat, no coercion. My life was comfortable. I just couldn't leave.

I remember visiting a zoo when I was young and seeing a few gorillas through a glass wall. A constant stream of humans gawked at them, but they seemed bored, sad and disinterested. Now I knew how they felt.

So I went through the motions. I woke up in my fake house, drank something that tasted like coffee, and watched the morning show. As well as the TV, my keepers allowed me real-time access to Earth's internet, and I did a little surfing. It was all one-way traffic, of course. Scrolling through the online version of the *Central Western Daily*, I found in it, a missing person's notice.

Police Seek Public Help In Locating Missing Orange Woman

Jane Ann Robinson was last seen arriving home last Tuesday, around 5:30 PM. Due to the mysterious circumstances of her disappearance, police hold grave fears for her safety. Anyone with information is urged to contact Orange police.

There was a photo of me looking slightly startled taken from my driver's licence. The cops would be baffled, alright. I could imagine their report. "No signs of disturbance or forced entry," it would state. "Handbag with phone, purse and keys found on dining table. Vehicle in garage." It would seem to the cops that I had vanished into thin air. Which wasn't far from the truth. In time, I would be just another statistic.

As the days rolled into weeks, I became bored, sad, and disinterested like my gorilla cousins. I'm not sure if it was this, or the fact I stopped breaking things, that brought about a sudden change in my circumstances.

One morning I discovered I had a backyard. I walked out onto something that looked like green grass and, above a high fence, I saw a patch of blue sky. Without thinking, I plunged my fingers into the grass and felt damp soil beneath. I didn't know if it was real or not. It didn't matter. For the first time in a long while, I smiled.

Oh, yes, they were clever bastards.

I spent hours on that square of grass, wondering what a few plants would look like. I asked for and received some seeds. Anything I wanted, they said. With the help of the Net, I learned how to grow things. I started small, with a couple of herbs and some veggies, then moved on to shrubs and indoor ferns. A tree would be next. I read somewhere that humans were nothing if not adaptable, so I adapted. I created a piece of Earth, a little Eden in my alien enclosure.

I've also started writing. Putting my thoughts down has helped me gain some perspective on things. If I ever get out of here, my journal will make for some interesting reading. Now I take each day as it comes. Lately, though, I've been feeling lonely. It would be nice to have someone to talk to other than a disembodied voice. Which got me wondering. Why grab only half of the species? I mean, here I am, a perfectly functional female, and well …

There was a knock at the door and I jumped, nearly spilling my coffee. This was new. No one had ever come to the door before. I felt a growing sense of unease about what lay on the other side. Keep moving forward, I told myself. I got up and opened the door. Standing on the welcome mat was a very confused looking man. He was tall and dark-haired, with just the right amount of vulnerability in his blue eyes.

"Where am I?" he asked.

He was the first human I had seen in almost a year. I hugged him fiercely.

Oh yes, they are clever bastards.

Garry Dean lives on the Mid North Coast of NSW and has been a fan of SF, ever since his older brother took him to see *2001 a Space Odyssey* for his eighth birthday. Although he was painting, and writing about other worlds during his teens, it wasn't until he was in his 40s that Garry began to write seriously. Hampered by a genetic eye disorder, he turned to adaptive technologies like voice recognition and text to speech. Garry's work has appeared in online magazines, including *AntipodeanSF*, *Quantum Muse* and *Daily Science Fiction*. <u>www.garrydean.wordpress.com</u>

Merrill Findlay

My Seven Sisters Dreaming

Big sky, inland plain, farmstead, veranda, and, across the creek, the fiery glow of Earth's sun sinking behind the ridge we call the Seven Sisters. If there's water in the creek, then it glows too, an ephemeral stream of liquid gold snaking through the twilight towards those seven conjoined hills on our horizon. And, slowly the stars, Moon and planets …

I love this view across our farm, its seasons, rhythms and moods. I especially love those Seven Sisters that burst so suddenly, so improbably, from our inland plain. I've gazed at them for more than half a century now and from every angle: through the front windows of the old house I grew up in, the veranda, the garden, the paddocks, shearing shed, creek bed, dam banks and stock route; from horse-back, ag-bikes, tractors, farm utes and cars, even from aircraft and Google Earth. I've walked them, climbed them, camped and picnicked beside them, even peed behind some of their biggest bushes. I know these Sisters intimately. They're part of me. And yet, paradoxically, I know so little about them: who were those seven siblings for whom the ridge was named, where did they live, what happened to them, and what did they mean to the clansfolk who lived here before my mob arrived with their cattle, sheep and ploughs? No one has been able to tell me, or not definitively. Why this silence?

Seven Sisters Ridge is in the heart of Wiradjuri Country. It entered whitefella records sometime in the nineteenth century when a colonial surveyor inscribed the name on a map. This 'foot soldier of empire', whoever he was, probably consulted the manager of the big pastoral station the ridge was then part of, probably even asked the Wiradjuri stockmen what they called it, and he might have documented some of the stories they told him too. But no records of these conversations, if they ever occurred, have survived as far as I know. All we have, therefore, is a ridge stripped of its stories, a toponym stripped of its provenance, seven mythic sisters stripped of their identities, a people stripped of their Country, and we settler-descendants left naked in a degraded environment of our own creation—yet surrounded by the material culture of those who were already here: charcoal from their campfires, tools they left behind, grooves in rock outcrops were they sharpened their axes, paintings on rock shelter walls, burials, bora rings, stone arrangements, dendroglyphs on the oldest remaining eucalypts, and hundreds of other tree scars where they removed slabs of bark for their shields, shelters, coolamons and canoes.

And the answers to my questions about those seven siblings and their namesake ridge? They're trapped, I suspect, in the cognitive gaps between the evidence Wiradjuri people left behind, the family memories the pastoral workers passed on to their descendants, and my mob's silences, occlusions, denials, confabulations and outright lies about what really happened on this plain when whitefellas arrived in the first half of the nineteenth century.

Now the Sisters are demanding their stories back. I can't ignore them, but it's not my place to 're-awaken' their Wiradjuri stories. All I can do is reach into my mob's history to recover some of our own 'forgotten' Seven Sisters stories. As I do so, I must recognise, however, that this ridge was almost certainly a sacred knowledge site and astronomical observatory on the dreaming track that traces the creation journey of seven ancestral women and one or more men who pursued them across continental Australia and beyond. I know this dreaming by its whitefella name, the Seven Sisters Songline. It's not a 'line', of course, nor a single track; more a memory cache of stories connecting the Sisters' terrestrial sites with their skyworld, including the nebulous star cluster I know by its ancient Greek name, the Pleiades, and by its more recent scientific label, Messier 45.

The songline encodes the data, knowledge, values, laws, customs and ways of understanding the cosmos that Australia's First Nations peoples developed over tens of thousands of years. For Wiradjuri people of this inland plain, the stars, the Ridge, the creeks, and other geographical features were mnemonic devices to ensure that this all-important knowledge was remembered and passed on through the generations. In Time-Before-Whitefellas, children born here would have imbibed Seven Sisters stories with their mother's milk. They would have grown up singing Seven Sisters stories, dancing them, weaving, sculpting, painting and drawing them and, when these children were old enough, they would have been initiated into the stories' deeper meanings. Such stories would have bound people spiritually, emotionally and intellectually to this inland plain, to Seven Sisters ridge, to the entire cosmos in ways that I, a settler descendant, can barely imagine.

This knowledge system has survived for millennia, but it has one major vulnerability. It depends on human memory. If the oral transmission of stories from one generation to the next is interrupted or blocked, if the ceremonies cease, if the language is lost or degraded and the songs are no longer sung, then thousands of years of knowledge can be erased within a single generation.

I was born into a very different knowledge system with very different modes of data storage and transmission, and yet my cultural heritage also includes Seven Sisters stories that are similar to the Wiradjuri stories. The same star cluster, which animated Wiradjuri people's imaginations, regulated their seasonal practices, and connected them conceptually to the rest of the cosmos did the same for my ancestors. My Seven Sisters stories take me back to priestess astronomers in the ziggurats of ancient Sumeria, in present-day Iraq, for example, and to stargazers in ancient Egypt, Anatolia, Persia, Arabia, India, the Aegean islands and beyond. I gaze in wonder now at the same celestial objects that inspired these ancestors, even though I don't necessarily believe the stories they told, except in a mythic way. The stories I internalise tend to be evidence-based ones that we lump together under the rubric of Science, including Geology and Astronomy, which is what I googled when I wanted to know more about Seven Sisters Ridge.

What my googling revealed, however, was the scale of my own ignorance, because I didn't know enough to interpret the scientific data I found! I rang a government geologist in Orange, our nearest big town, and asked for help. He knew Seven Sisters Ridge well. Yes, part of the Byong Volcanics, he said. A stratigraphic unit formed some 420 million years ago when viscous lava leaked from fissures in an ocean floor somewhere off the coast of the now-extinct super-continent Gondwanaland. He seemed delighted that a member of the public would be interested enough to ask him about it!

The lava cooled into rhyolite, the pink and grey crystalline rock at the core of each of the seven peaks, he continued. Millions more years of sedimentation, erosion, compaction, distortion and evolution later, a couple of climate changes and mass extinctions, more rifting and continental drifting, another burst of plate tectonics as Gondwanaland broke up and new continents and oceans formed, more heating, cooling, faulting and folding, yet more sedimentation, compression and erosion—and lo, a ridge of seven crystalline peaks on a vast inland plain hundreds of kilometres from any ocean.

What this geologist gave me was another dreaming, a creation story as read from the rocks themselves, which allows me to see Seven Sisters Ridge through the lens of nearly half a billion years of our planet's geological evolution. I've since learnt that the Sisters' lava leaked from their submarine fissures in a period geologists call the Silurian, a thirty-million-year sequence of terra-forming events identified by one of nineteenth-century Britain's most prominent gentleman geologists, Sir Roderick Impey Murchison, and his wife Lady Charlotte, a skilled amateur geologist in her own right.

In 1831, while British 'squatters' were claiming vast tracts of inland New South Wales as grazing land for their cattle and sheep, the Murchisons embarked on a grand carriage tour of old South Wales to collect fossils and study rock formations. They took with them one of their maids, 'two good grey nags, and a couple of extra saddles for day-trips', a Romantic European sensibility, and a very Scottish Enlightenment passion for classifying things and imposing order upon the world. What they found inspired one of Geology's most consequential insights: that fossils in sedimentary rocks in one place could be used to correlate the stratigraphy (and age) of rocks in other places. And, yes, the fossilised trilobites and brachiopods they found in old South Wales were the same or very similar to the fossils in the sediments of that now-extinct seafloor beneath Seven Sisters Ridge on the other side of our planet.

The Murchisons named 'their' stratigraphic sequence for the Britons who lived in south Wales at the time of the Roman conquest, a people we now remember by the name their conquerors gave them, the Silures. Sir Roderick presented his new geological classification to the British Association for the Advancement of Science in a paper he co-authored with a fellow geologist, the Reverend Adam Sedgwick, and further developed his ideas in a monograph he called The Silurian System. By this time, however, Sedgwick and Murchison were fighting a very public border brawl about where, in the emerging Geological Timescale, the Silurian period began and ended. Their spat was resolved decades later by a much younger fossil sleuth, Charles Lapworth, who added his own geological period to the timescale. He named his period for the Silures' northern neighbours, the Ordovicians.

Geologists and palaeontologists now agree that the Ordovician period began around 489 million years ago with an evolutionary burst of complex life forms called the Great Ordovician Biodiversification Event (GOBE). Intriguingly, the GOBE coincided with an extended period of intense meteorite bombardment after a massive asteroid broke up within our solar system. So, did that celestial meteorite storm contribute to, or perhaps even trigger, the rapid proliferation of new life forms on Planet Earth in the mid-Ordovician period?

And what about the Ice Age that killed off an estimated 85 per cent of species at the end of the Ordovician? Was that mega-extinction event induced by extreme volcanic activity, as some scientists suggest, or by the too-rapid movement of our planet's continental plates, or by wobbles in Earth's tilt, or by eccentricities in our planet's orbit around the sun, as other scientists have hypothesised? Or was it driven by intergalactic forces? A too-close encounter with a super-cloud of galactic dust, or a burst of gamma radiation from a relatively close hypernova, as astronomers have suggested?

Ah, yet more questions I can't answer! All I know is that right now, hundreds of millions of years after that almost complete annihilation of life on Planet Earth, this sentient being, a very distant relative of the life forms that survived the Ordovician-Silurian cataclysm, is tap-tap-tapping away at her laptop in a café in a small country town near our farm searching for words to describe the cosmic improbability that she—and you, dear reader—exist at all and are able to interrogate the universe in this way!

But back to those ancient Britons the Romans called the Silures and Ordovices: what stories did they tell about the cosmos and their place in it? Unfortunately, we can't know because theirs was an oral culture too. The little we know about them has been gleaned from artefacts they left behind, landscapes they created, their skeletal remains, DNA and other molecular evidence, from folklore, traditions and toponyms associated with them, from the Celtic languages some of their descendants still speak, and, of course, from the writings of their Roman conquerors, including the imperial historian Gaius Cornelius Tacitus, whom I first met in my schoolgirl Latin classes.

These sources tell us that, at the time of the Roman invasion, ancient Britons lived in hierarchical farming communities led by a noble warrior class and a clergy of Druids. We also know that the Seven Sisters of the Pleiades were as important to them as they were to Wiradjuri people. The Pleiades told Silurian and Ordovician farmers when to sow and harvest their crops; they told their seafarers when they could sail safely; and they told the Druids when to conduct their religious rituals and festivals. When the Pleiades reached their highest point in the sky, their culmination, ancient Britons knew it was time to celebrate Samhain, the harvest festival, for example. They also apparently believed that the veil separating the living from the spirits of their dead was thin enough for them to commune with their ancestors at this time of year. The Roman Church appropriated this date as All Saints Day or Hallowmas. Today we call this festival Halloween—and forget its celestial origins.

Like the Wiradjuri, the Silures resisted the invasion and colonisation of their homeland. Indeed, they weren't 'pacified' until around 77 CE when Tacitus's father-in-law, Gnaeus Julius Agricola who, by then, was governor of the Roman province of Britannia, led a legion across the Mennai Strait to attack the Druid's stronghold of Ynys Môn, the island of Mona or Anglesey, in present-day Wales. Agricola's army destroyed and defiled the sacred sites and massacred the Druids. They also killed the Britons who were defending the island, and the refugees, including women and children, who had sought sanctuary there. According to Tacitus, the Druids and the Ordovices were exterminated. Any

Silures who survived were resettled or 'concentrated' in a purpose-built town the Romans called Venta Silurum near Isca Augusta, the Second Augusta legion's fortress in South Wales.

Some sixteen hundred years after the Roman legions withdrew from Britannia this history was grimly repeated when some of the descendants of those ancient Britons' did to the Wiradjuri what the Romans had done to their ancestors. An all too familiar story: invasion, conquest, colonisation, dispossession, subjugation, oppression and discrimination followed by denial and 'forgetting'. I've heard it told from so many different perspectives and my search for answers to those Seven Sisters questions has revealed to me even more!

Because I learned that one of the men responsible for setting in motion the cataclysmic events that decimated the Wiradjuri nation and stopped or interrupted the intergenerational transfer of knowledge, including Seven Sisters stories, was, ironically, an astronomer! A privileged member of Scotland's landed gentry, a son of the Scottish Enlightenment, a veteran of the Napoleonic Wars, a devout Protestant Christian, and a fellow of the Royal Society of London: Sir Thomas Brisbane, the major general who, in 1822, became the sixth governor of the colony of New South Wales and, according to some, 'the Father of Australian Science'.

Brisbane didn't want to govern a penal colony at the farthest reach of the British Empire, though. All he wanted to do was confirm the shape of our planet (was it a lumpy sphere or pear-shaped?) by measuring a southern meridian and be the first European to catalogue the stars of the southern sky. For Brisbane, the governorship of New South Wales was simply a means to these ends. He disembarked at Sydney Cove in November 1821 with his much younger wife, Lady Anna Maria, and their infant daughter, Isabella. He also brought, at his own expense, two personal staff, astronomer Carl Ludwig Rümker and technician James Dunlop, and all the kit he needed to build and operate an astronomical observatory.

The new governor and his entourage settled into old Government House in Parramatta. Work began on the observatory almost immediately. Six months later, in May 1822, Brisbane, Rümker and Dunlop began observing and documenting the southern stars and other celestial objects. But Brisbane also had that pesky day job to attend to, the administration of the colony and implementation of the British government's new policies, including the expansion of the pastoral industry west of the Blue Mountains.

Brisbane's predecessor, Lachlan Macquarie, had commissioned the first whitefella track across the mountains into Wiradjuri Country a few years earlier but had restricted the number of colonists and convicts who used it. Sir Thomas Brisbane did not. For Wiradjuri people, and the environment they were part of, the consequences of this invasion of foreigners and their livestock were brutal. When the impacts became intolerable, they launched a guerrilla campaign to defend and liberate their Country as the Silures and Ordovicians had done in Britain. Brisbane retaliated with a declaration of martial law.

Years later, a grandson of one of the beneficiaries of Brisbane's pastoral expansion into Wiradjuri Country, William Henry Suttor, recounted some of his own family memories from this era. Suttor's father, also called William, had been overseer on his father's pastoral station near Bathurst in the 1820s and had befriended local Wiradjuri people, including their resistance leader, Windradyne. Suttor Snr had thus personally witnessed the consequences of Martial Law for Wiradjuri people and had passed these stories on to his children. In recounting them, William Jnr evoked Tacitus's account of Agricola's campaigns against the Caledonians, Brisbane's Celtic ancestors in what is now Scotland. "When martial law had run its course, extermination is the word that most aptly describes the result," Suttor wrote. "As the old Roman said, 'they made a solitude and called it peace'."

That 'old Roman' gave these words to Calgacus, a leader of the Caledonians' military resistance: Auferre, trucidare, rapere, falsis nominibus imperium; atque, ubi solitudinem faciunt, pacem appellant. "To robbery, slaughter, plunder, they give the lying name of empire: they make a solitude and call it peace." Solitudinem can also be translated as 'desert' or 'wasteland', which, for the Wiradjuri who survived Brisbane's war and pastoral expansion, was an apt description of what they experienced. The horrors, the traumas, the attempted extermination that Brisbane unleashed upon Wiradjuri people still haunt their descendants today.

That view across our farm continues to move me, but now I see Seven Sisters Ridge and that cluster of stars very differently: as through the lens of Sir Thomas Brisbane's telescope. And yes, there are ghosts, too many of them, but, in the distance, I can sometimes also glimpse the future. Wiradjuri people restoring their 'lost' stories, reviving their 'forgotten' science and ceremonies, and re-connecting the Ridge to other sacred knowledge sites along the Seven Sisters Songline. I can also see we other Inlanders drawing sustenance from the Sisters' peaks and their companion stars. Because, no matter how conflicted that view is now, no matter how painful, it reminds us that our pasts, presents and possible futures are far more entangled than any single story can tell and that how we see the cosmos depends on where and when we view it from, and what lenses we use.

My thanks to geologist Gary Burton from the Geological Survey of NSW office in Orange, NSW, who explained the geological history of Seven Sisters Ridge to me, and to the staff of Geoscience Australia who gave me the links to relevant geological maps and other references; to my High School Latin teacher who introduced me to Tacitus; and to the Wiradjuri people who've educated me about the ongoing intergenerational impacts of colonisation in their communities. All misinterpretations and inaccuracies are my own, though, of course.

Sources used include

Barash, M.S., 2014, Mass extinction of the marine biota at the Ordovician-Silurian transition due to environmental changes. *Oceanology* 54 (6), 780-787.

Berger, A. and Q. Yin, 2012. "Modelling the Past and Future Interglacials in Response to Astronomical and Greenhouse Gas Forcing" in *The Future of the World's Climate* (2nd ed.), edited by A. Henderson-Sellers and K. McGuffie, 437-462. Boston: Elsevier Science.

Bhathal, R., 2012. "Some Scientific aspects of Parramatta Observatory: Symposium – Commemorating Governor Sir Thomas Brisbane" in *Journal and Proceedings of the Royal Society of New South Wales*, 145 & 446: 111-127.

Condie, K.C., 2011. "Great Events in Earth History" in *Earth as an Evolving Planetary System* (2nd ed.). K. C. Condie (ed), 357-435. Boston: Academic Press.

Geoscience Australia, Australian Statagraphic Units Database, Byong Volcanics. http://dbforms.ga.gov.au/pls/www/geodx.strat_units. sch_full?wher=stratno=3226, accessed 6 July 2018.

Kelly, L., 2016. *The Memory Code*, Sydney: Allen & Unwin.

Liston, C., 2012. "Sir Thomas Brisbane-a man of scientific method" in *Journal and Proceedings of the Royal Society of New South Wales*, Sydney: Royal Society of New South Wales.

Makdougall Brisbane, T. and W. Tasker (eds.), 1860. *Reminiscences of General Sir Thomas Makdougall Brisbane*. Edinburgh: Thomas Constable.

Brewer, Richard J., 2002. *Birthday of the Eagle: The Second Augustan Legion and the Roman Military Machine, Amgueddfeydd ac Orielau Cenedlaethol Cymru*, Cardiff: National Museums & Galleries of Wales.

Melott, A. L. and B. C. Thomas, 2009. "Late Ordovician geographic patterns of extinction compared with simulations of astrophysical ionizing radiation damage" in *Paleobiology* 35(3): 311-320.

Murchinson, R. I., 1839. The Silurian System. London: James Murray.

O'Donoghue, J., 2016. "The second coming" in *Origin, Evolution, Extinction: The epic story of life on Earth. New Scientist: The Collection* 3(2).

Pohl, A., et al., 2016. "Glacial onset predated Late Ordovician climate cooling" in *Paleoceanography* 31(6).

Sherwin, L., 1980. "Faunal Correlation of the Siluro-Devonian units, Mineral Hill-Trundle-Peak Hill Area" in *Quarterly Notes: Geological Survey of New South Wales* 39: 1-13.

Suttor, W. H., 1887. *Australian stories retold and sketches of country life*. Bathurst: Glyndwr Whalan.

Tacitus, G. C., 1999 [c. 98 CE], A. Church and W.J. Brodribb (trs), *Life of Gnaeus Julius Agricola (The Agricola)*, Internet Ancient History Sourcebook, New York City: Fordham University.

Tyler, P. J., 2012. Sir Thomas Brisbane-Patron of Colonial Science in *Journal and Proceedings of the Royal Society of New South Wales*, Sydney: Royal Society of New South Wales.

Merrill Findlay is a writer and cultural development practitioner who now lives in Forbes, NSW. Her published work includes a critically acclaimed novel, *Republic of Women* (UQP 1999), book chapters, blog posts, speeches, scholarly articles, conference papers, an opera libretto, and features for the mainstream press, such as *The Age*, *Good Weekend*, and the *Canberra Times*. Her most recent cultural interventions include The Skywriters Project and the Inland Astro-Trail. www.merrillfindlay.com

Martha Morrison Gelin

My List

1947

Number one on my list is this:
1. Own my own *Encyclopedia Britannica*

Number two is this:
2. See Halley's comet. I know, I know. It's a long way away, and I'll be old, probably 48 or 49 when it comes back, but I plan to see it.

My mother thinks I'm too young to have a list, but I'm already 10, and anyway she told me her list for my life.

Her list:
1. Get your chin fixed (She thinks it's too small)
2. Get your front teeth straightened (Well, that is OK. I agree with that one)
3. Get married (Huh!) (And I know that means have kids, too)
4. Make sure you go to college (I agree)
5. Keep your room tidy
6. Don't try to vacuum anymore while you are reading a book (She actually took my list away from me and WROTE that on it!)

So I'm not talking to anybody any more about my list, but this is the rest of it.

3. Write a book

4. See the book in bookstores and libraries and other people reading it

5. See a really good Northern Lights

6. See a total solar eclipse

7. Get a PhD

8. And I really don't think this will ever happen, but SEE A SUPERNOVA!

I'm not sure the list is finished, but those are things I want to do. I'm not too sure about the PhD thing, but I know if you get one you get to be called "Doctor" and I'd like to be Dr Morrison or, to friends, Dr Martha. But I don't want to be a real doctor.

I've been reading about stars and planets and thinking about how big and how far away they are, maybe too far for us Earthlings. I read that name, too, in a story, and I like it a lot better than "German" or "American" or "Japanese". I think everybody should just call themselves an Earthling. And I'd really, really, really like to see a supernova.

1987

I don't think this cloud cover will end soon enough to see 1987A. They say the supernova is visible in the Large Magellanic Cloud, but we can't see it from Bathurst—yet. Come on Mother Nature. Clear up this cloud! I can't stand it!

And She did! There it is!! Gee, it's so clear. It's smaller than I hoped, but it's a bright dot where no bright dot ever was before. So, forty years after writing my list, I've done the one thing on it I thought I'd never get to do! And I've also done most of the rest.*

2019

I still think "Earthling" is the best name for all of us. And I missed the last total solar eclipse!

Notes

* #s 1, 2, 3, 4, 7, 8
NB: #5 – Did see one smallish, very green, hand-shaped one in Ontario.
      #6 – Several partial solar eclipses.
NB2: From my mother's list: #s 2, 3, 4. Plus kids. Did try to vacuum while reading, but it never really worked.

MoonDitty: Concerning the Moon and the Media

Super Moon! Blood Moon! Wolf Moon! Wow!!
I guess the world is ending now.
And from what the headlines say, it's either next week or later today!

Blue Moon! Full Moon! Lunatics loose!
Danger, or beat-up? You deduce.
Eclipse madness, or is it a high? Push the right buttons, and folk will buy.

Silvery Moon. Moonbeams. New Moon. Wishes.
Songs and stories; cows and dishes.
Walks in Moonlight; ladders on sea. Kisses, romance, baby makes three.

Harvest Moon. Hunting Moon. First Moon. Seasons.
Holidays, holy days, marking out the reasons.
Folks need guidance, wouldn't you say? And we have the media.
Big Hooray??

Moon Rocks! Moon Walks! Far-Side Lander!
Moonlit scientists. Awe-struck bystander.
Man in the Moon. Rabbit. Goddess. Cheese.
We should just enjoy it. Cut the hype, please!

Sky As Landscape

The Central West of New South Wales has always had a familiar feel to this immigrant because it reflects the general geography of my home state of Tennessee—altitude in the east, with smoky blue heights; pleasant plateau with rolling hills in the middle; gradual sloping down to western flatlands. But Tennessee has a lot of fresh water—very different indeed to our NSW Central West. The Macquarie River, running through my permanent home town of Bathurst, would be called a creek back where I grew up.

'Home' has moved around for me over the decades. At various times it has included southern, northeastern and western USA, northern Ontario in Canada, and Bolivia and Colombia in South America. My first home in Australia was Launceston, Tasmania, and, for the last thirty-five years, Bathurst. But always and everywhere, the sky has been part of my landscape. The daytime sky often provided me with stability—an 'Oh, hello Cumulus. How're ya goin'?' kind of feeling while I was getting my feet settled into new ground and getting familiar with new light, vegetation and human structures.

There are many different hues and shades of blue sky, but, wherever I am, the clouds have retained their same familiar shapes. I did get a fright on my first visit to the western part of the Central West, though. The sky settled lower and turned a dirty yellow-greenish brown as I was driving. In Tennessee, such conditions signal tornado weather, so I hit the accelerator to get to the next town. "It's just the dust!" a local told me there. "Wind's up today."

Then there's the night sky. Same, same, but different. I find that darkness everywhere feels the same, even when the heavenly bodies look different. To me, the strangest difference north to south is the face of the Moon. I'm always glad to get back to our Southern Cheerful Face (I generally ignore the rabbit). The lugubrious Northern Hemisphere Man looks a bit depressing by comparison.

Here in Bathurst, as in my rural Tennessee childhood, the stars are a real presence, even from my garden, a fifteen-minute walk from the lights of Bathurst's CBD. But I cheat. My back garden is tiny and tightly enclosed by the rear wall of our two-storey house, a neighbour's shed, the retaining wall, fence and shrubs of another neighbour, and by my water tank and another neighbour's fence. That gives me a dark canyon from which I can see the stars of the southwest slice of sky. In the front garden, where tall bottlebrush trees screen the downtown lights, I have a banana chair view of the northeastern and eastern skies, including their regular meteor showers.

I don't know how anyone can ignore the sky: all that whipped cream piling up of grey, silver and lavenderish storm clouds, the ever shifting shades and shapes of white clouds, the daytime phantom Moon—let alone the stars. How it is that some folk never look up? I've found, however, that if a sky phenomenon is mentioned on social media, then everyone's talking about it, and people turn out in masses to see it. They expect something spectacular, I assume. Like a Blue Moon that is literally bright blue!

Sunset is one of my favourites: the ever-changing warm colours, the lengthening shadows, the fade of visibility, the turn towards night. In my childhood, the twilight meant dinner time, time to stop the games and come inside. Sometimes, in the deepest part of summer, we had enough light after dinner to get back to our games—Hide and Seek, Red Rover, Slingin' Statues—or to just sit and talk: 'Move over. You're pushing me off the step!' 'Can I have a bite of your cookie?' 'You're sooo lucky to have a dog.' 'Oh! First star! I get to make a wish.'

Sometimes, after a family outing, we'd drive home in the dark. I'd always lay my head against the car window and watch how the Moon moved with us. If it was close to Moonrise, I'd sometimes see the Moon get bigger or smaller, and I'd think of magic. My father said it was because when the rising Moon was close to trees or buildings, it looked bigger. 'Just watch. It'll look its right size when it's higher.' He was always right, but I liked the magic idea better. Perhaps some good fairy or dark witch playing with the Moon.

Sometimes my father took me and my little brother outside to point out the constellations. I got to know the Big Dipper, the Little Dipper and the North Star. My father was a bit of an outdoorsman and thought everyone should know how to find North. I could usually find Orion and Cassiopeia, too, but, in those days, the Great Bear was my favourite. To me, this constellations has always been a Bear rather than a Dipper.

One of my summertime pleasures was seeing how a sunbeam could light up the sandy and gravelly bottom of the shallows in Stones River where we kids played and swam. The sunlight heightened the colours of the leaves and flowers and made the white clouds shine against the powdery-blue sky. The sparkles on the rippling surface of the water reminded me of stars.

Last time I visited Stones River, I stood at the edge of the bluff where we used to climb down to the river. It was a bluff no longer, but part of a shoreline. Stones River, its deeps and its shallows, and part of the community I had lived in, were somewhere over a hundred feet below me, under the surface of a huge, dammed lake. But the same stars were still sparkling on the water's surface.

In 1945, just after WWII, when I was eight, we moved to Cochabamba in Bolivia: Southern Hemisphere, stronger sun, more sunburn, and no Great Bear nor North Star. Instead, we had the Southern Cross. My father showed it to us and made sure we knew how to find South.

Years later, in the 1970s, when I lived in northern Canada, Orion became my favourite celestial pal. There he was looming over the horizon and striding across the night, a kind of lodestone when the world went stupid. The back steps of my house became my contemplation place; the revolving constellations and planets were as much a part of my being as the wonderous snow formations and new growth that leapt from under the melting ice in Spring. Then back to the USA and city lights: fewer stars, lots of red and green lights. I moved on to the Nevada desert, with its hard, tile-blue sky, storm clouds cresting and breaking in gigantic waves over the Sierras, and astonishing red, orange and purple sunsets. The stars struggled against the Vegas lights.

In 1981, we moved from near Las Vegas to Launceston in Tasmania. I brought on my first celestial shock when I asked to be taken outside in the dark and re-introduced to the southern night sky. I found the Southern Cross and was delighted to see that it still hung sideways. My host pointed out The Saucepan. What? Wait! I turned around, bent backwards as far as I could. Just as I thought! It was Orion! But he was upside down! I was so happy to see my old pal, I decided to accept his strange orientation. The Bull and the Seven Sisters seemed OK in their new positions, but Sirius had evidently learned how to fly.

My second shock occurred a few evenings later while I watched the sunset from our western window, glorying in the opalescent Tasmanian light and the lengthening shadows on the tree covered hillside. Suddenly, I went dizzy and disoriented and had to grab the window sill. A few moments later, I was fine. The next day I sought our neighbour, a geographer. 'Does the sunlight fall at a different angle at sunset here from the western USA?' I asked.

'It does, indeed,' he told me.

That evening I watched the sunset again from our western window and waited for the dizzy spell. It never came. Nor did it when I tried for it the first time I visited back in the USA. One acclimates very quickly to different angles of sunlight, it seems.

How have I reconciled myself to my Orion of the South? By seeing him at play—the Hunter cartwheeling across the sky, joy in every line. I welcome him back each November, and I greet him every time I catch the Bathurst Bullet in the pre-dawn winter. Fewer constellations present themselves to me now, but Venus blazes most amazingly before some dawns. And I do like the curve of the Scorpion's tail.

When we travel to the northern edge of Oz, as I hope to do soon, I'm going to seek the darkness to see which old pals from my former homes are lingering around the horizon.

Martha Morrison Gelin was born in Tennessee, USA, more than eighty years ago. She and her family have lived in many different places since then and now call Bathurst home. Martha worked professionally in human services, academia and private counselling practice, and has written extensively on Sudden Infant Death Syndrome (SIDS), meditation, research evaluation and sexuality, including two books for parents about talking with their children about sexuality. Now in 'retirement', she wants to write more creatively about some of her many other interests— including supernovae!

Ian Gibbins

No Vacancy

We are looking for landing space.
So far away from home base, we
seek rest, shelter from each other,
from the incessant silence around
our insulated pods, our antennae
straining for familiarity, a clue.

Refreshed by your solar wind, we
trust your civic pride, we have faith
in your management of new arrivals.
We will access your local datasets,
scan your facilities, their files, but
now we ask your thoughtful advice.

Could we arrange to meet you at
The Railway Bowling Club? Dine
with you at The Starlight Lounge?
Should we pre-book rooms at The
Bushman's, Parkview, The Apollo?
Can any accommodate our requests?

We have mapped Clarinda, tracked
Forbes, the Newell, and beyond, yet
we struggle for understanding, the gist
of who you are. We must soon decide
on descent or farewell, greetings or
departure in clouds of cometary dust.

Give us a sign. Send us a message.
Our decoders are primed and ready.
We hope you have not missed us,

our passage unobserved, our signals
lost in traffic, in your rush towards
tomorrow. Just stop. Welcome us.

Ian Gibbins is a widely-published poet, video artist and electronic musician with four collections of poetry: *Urban Biology* (2012); *The Microscope Project: How Things Work* (2014); *Floribunda* (2015) and *A Skeleton of Desire* (2019). Ian's videos have been shown at international festivals, having won or been short-listed for multiple awards. His audio and video work has been commissioned for public art programs and has featured in several gallery shows and installations. In his former life, Ian was an internationally recognised neuroscientist and Professor of Anatomy at Flinders University, South Australia. www.iangibbins.com.au

Emma Gibson

Face full of stars

Afar, almost antipodean
(almost; the true antipodes from home would plunge me seawards
To peril, to sharks)
I prepared to leave the last port of my odyssey
To find my way home.

Returning, I saw for the first time what I had left:
The slender beauty of silvery gums
Distant blue hills
The yellow of wattle and canola
Sunburned skies turning the landscape purple as the sunset
The clear view of the stars.

I remember I envied my best friend Garth's face full of freckles
Like the Milky Way
A galaxy mapped across his face.
I wished for the same

Half smiling, my mother promised
My time would come.
They have never faded, not like his
Each year, they darken in the sun.

On a dark clear night in the country
I craned my neck to watch each star pop out.
The Milky Way is bright
From the lawn at my Mum's.

I returned to a life stuck in cycles from the past: not a repeat,
but an echo.
Strangest of all, I found myself living with Garth.

His freckles have faded now;
I've given up trying to cover mine.
They are a stamp on me of where I come from
This place large under the sun.

I walk the same routes, as though I was never gone
My body still remembers the curve of bus as it ferries me home.
In our backyard, I stand next to the Hill's Hoist and look up
It's not the bright lights of dark nights in the bush
But they are the same stars
Millions of unspent wishes.

Emma Gibson is a writer, theatre-maker and poet with a focus on place-based work. Her work has been produced internationally, including in Iceland, Sweden, the USA and the UK. Her poetry and prose has been published in *Iceview*, the *Skagastrond Review* (Iceland), *Broadway Baby* (UK), *LIP Mag*, *BMA Magazine*, *Seizure* (Australia) and in several anthologies. Her most recent theatrical work was *Tourmaline,* adapted from the Randolph Stow novel at the Street Theatre in Canberra. Emma's previous plays include *Bloodletting*, *War Stories* (as co-writer), *Johnny Castellano is Mine, The Pyjama Girl, Widowbird* and *Love Cupboard*. Emma is currently based in Melbourne.

Suzie Gibson

The Dish and the Big Skies of the Central West

The 1969 Moon landing was a triumphant moment of the 20th century. It realised young President Kennedy's 1961 declaration that within the decade humans would land upon the Moon and then return safely to Earth. It was also a globally televised achievement, people across the world witnessing Neil Armstrong's first-ever lunar walk. Much like the impact of Kennedy's assassination six-years earlier, individual and collective memory oscillates around this event. It is one of the most lauded achievements of humanity to date.

Over the ensuing decades, countless documentaries, films and songs have celebrated the Moon landing. Al Reinhart's documentary film *For All Mankind* (1989) meticulously chronicles NASA's Apollo mission, while Mark Cowan's *Magnificent Desolation: Walking on the Moon* (2005) provides a comprehensive account of NASA's entire Apollo program. And then, of course, there are mnemonically powerful pop songs with at least tangential links to the Moon landing that helps keep the memory alive—the Police's "Walking on the Moon" (1979) and R.E.M.'s "Man on the Moon" (1992) immediately come to mind. And Brian Eno's album *Apollo: Atmosphere's and Soundtracks*—originally intended as an accompaniment to Reinhart's film—beautifully captures the mood of the lunar landscape.

More recently, conspiracy theories have accumulated around this significant event. French mockumentary *The Dark Side of the Moon* (2002) has contributed to a plethora of YouTube speculations about the Moon landing being some kind of hoax. Such efforts to erode its facticity are largely expressed through the echo chambers of the internet—a sphere barely imaginable when the Moon landing took place. It is also notable that most of the speculation, celebration and commemoration has been doggedly USA-focussed.

Such an American perspective tends to obfuscate other voices and points of view.

In light of this, there is a unique Australian story to tell about this epoch-defining moment. In this antipodean scenario, the big skies of Central Western New South Wales play a significant role. Rob Sitch's *The Dish* (2000) tells a compelling yarn about Australia's regional involvement in the Moon landing. In doing so, his film showcases the spectacular big skies of Parkes in the Central West. At the time of the Moon landing, the Parkes Observatory was the largest radio telescope in the world, and it was instrumental in receiving signals during the Apollo 11 mission. It is quite extraordinary to think that regional New South Wales played such a noteworthy part in America's drive toward the Moon.

The Dish also pays homage to the various individuals who toiled inside the monumental radio telescope—Cliff Buxton (expertly played by Sam Neill); Al Burnett (a NASA technician played by Patrick Warburton); Glenn Latham (delightfully played by Tom Long) and Ross 'Mitch' Mitchell (dryly played by Kevin Harrington). Such likeable characters remind us of the human element in all of this, an aspect that can be easily lost to science and technology.

The Dish is an important Australian film for many reasons, not least for its acknowledgement of Australia's involvement in such a crucial moment in world history. The film's breathtaking cinematography also reminds us of Australia's astonishing and unique rural landscapes, not least the vast dome of sky that epitomises our outback. The pristine night skies of Central West New South Wales might indeed be seen as another character in this film. Certainly, the location was a key contributing factor in the Moon landing: if not for this remote outpost Armstrong's iconic lunar stroll might not have been possible.

Setting the Scene

The Dish (2000) begins with the sweeping plains and skies of Central Western New South Wales. Set in the regional town of Parkes, *The Dish* reveals the unique landscape of regional Australia. In this opening scene, the camera traces the journey of an aged Cliff

Buxton who drives through an autumnal sun-drenched Central West in search of Parkes' gargantuan radio telescope. When he arrives, the Dish emerges as an extraordinary structure dwarfing the aged body of Buxton. As Buxton contemplates this mighty observatory, a montage sequence of historical footage is interpolated—we see black-and-white-images-of-Kennedy's famed 1961 declaration and colour scenes of astronauts and scientists. The distance and difference between tranquil Parkes and the American scene could not be starker. This scene also reinforces Australia's geographical location as a world far removed. However, such remoteness was crucial to the success of its radio communication.

Through Buxton's memories we are transported back to 1969—a scene of great excitement and anticipation for the town of Parkes and the surrounding district, as we witness locals eagerly preparing for the Moon landing. Enthusiastic shop-owners and local identities are seen gathering the community together as they prime themselves for an event that will change their small the town and the wider world forever.

Like many Australian films—Sue Brooks's hilarious *Road to Nhill* (1997) and Rob Sitch's earlier *The Castle* (1997)—*The Dish* lovingly chronicles the lives of ordinary folk. There is also a distinctively Australian, dry and laconic style of humour on display. Part of the comedy is situational as Parkes's Observatory, the largest receiving station in the world at that time, is in the middle of a sheep paddock. Such an oddity, even absurdity, is disarmingly hilarious. Graeme Wood's clever cinematography often captures the Dish amidst a flock of ewes, humorously undercutting the technological sophistication.

The film's amusing tone is compounded further by the self-effacing crew working inside the Dish: Ross Mitchell, the Dish's movement controller; Glen Latham, the telescope's electronics man; Al Burnett, NASA's signal communicator; and finally, Cliff Buxton, colloquially known as 'the Dish master'. A young enthusiastic journalist (played by Neil Pigot) light-heartedly asks this group:

> When you think about it, the Americans spend 10 years, billions of dollars to let us watch a man on the Moon and in the end it all falls to you blokes ... I mean how do you feel about that?

The ocker character Mitch quickly retorts: "A lot better before you opened your trap!" Attempting to preserve the dignity of the crew, the NASA man solemnly responds: "We feel confident that we have the expertise to complete our role."

The idea of a remote group of Australians playing an instrumental role in the Moon landing is at the core of the film. In another scene compounding the significance of Australia's contribution, we see the Prime Minister at the time—John Gorton, played by Billie Brown—make a rallying parliamentary speech: "And so Australia finds itself as a vital cog in this grand endeavour … And it will be one of the proudest moments of Australia's scientific history!"

At the end of his dramatic oratory, we learn that Gorton will visit the town of Parkes to honour its scientific contribution. Cleverly, we learn this through a deft scene change as the camera leaves the realm of national politics to focus upon Parkes's local mayor Bob McIntyre (played by Roy Billing), who is seen rejoicing over the Prime Minister's visit. 'Bob', the local Mayor, also takes this opportunity to congratulate himself over his own far-sightedness in pushing for the construction of the radio telescope. Photographs and framed images of the dish adorn his office. He sees the radio telescope as the crowning glory of his political career.

The self-interest and importance of this local mayor is humorously contrasted with a celestial aerial shot that captures the immense telescope from above. The Parkes Observatory is beautifully juxtaposed against the lush grasslands, paddocks and trees of the Central West. As the camera traces the Dish's unique engineering details, radio communication sounds remind us of its grand technological purpose. We also get a sense of a vast expanse beyond the world of self-interested politics and politicians. In fact, there is a clever dialectic or a dialogue that goes on this film, distinguishing the mortal sphere of human politics from the eternal realm of space.

The Big Skies of the Southern Hemisphere

When gazing upon the night skies of the Central West, one can quickly be moved to imagine worlds and galaxies beyond the sphere of our mortal existence. There is also something quite transcendent and profound about this experience of gazing upon infinity. An overwhelming sense of remoteness can take hold. Perhaps this is particular to the Antipodean experience of night skies, because the international focus tends to be tilted toward the lives of those living in the Northern Hemisphere. However, while being 'down under' might be geographically lonely, it nonetheless provides a special kind of isolation that was crucial to tracing the passage of the Apollo 11 spacecraft.

The Dish dramatically brings this theme of remoteness to the fore in a scene that dramatises American Al Burnett's concern over a change in the NASA "figures". 'Dish master' Buxton seeks to address the anomaly and calls upon Glenn, the Dish's radio signals officer, to explain. Glenn simply says that the figures were "wrong". Confused by the answer, NASA-man-Burnett and Buxton ask Glenn again why he changed the figures. Glenn reveals that Houston's calculations would make sense were they situated in the Northern Hemisphere. He offers to "change them back", noting that they'd be all wrong. This scene serves as a powerful reminder of how Australia's geographical and astronomical "difference" can also be experienced as a cultural phenomenon.

In *The Dish*, cultural differences between the USA and Australia are treated with a great deal of humour. This is wittily expressed in a scene with the village gossip as she informs Mayor 'Bill' of the NASA man's penchant for "pretzels". We also learn of the American Ambassador's plan to visit Parkes, which inspires locals to make a good impression. Preparations include enlisting the musical expertise of a local boy band tasked to play the American national anthem. They mistakenly play the theme music to the popular 1960s cop show, *Hawaii 5-0*, which is indicative of the powerful influence of US culture upon Australia.

Further cultural differences are comically explored as the Australian crew plays a game of cricket on the Dish's circular surface. During this game, ocker Mitch ruminates over their "professional unit" and how it is comparable to Houston Control. Scenes inside *The Dish* certainly testify to the professionalism of a team who are all deeply committed to the success of the Apollo mission. Strategically, *The Dish* splices actual footage from Houston Control, where we see countless be-speckled men leaning over lines of computers. The Houston footage sharply contrasts with Parkes's towering dish, which is often photographed against a big night sky. Remoteness and distance figure again as powerful tropes as the film counts down to the dramatic moment of the lunar landing.

Not all is smooth sailing, however. A blackout causes the Dish to lose its power source, wiping all data from its computers. In this scene, the remoteness of the all-encompassing dark night sky takes over. This temporary disaster suspends cultural differences as the NASA man Al and Buxton's team restore the coordinates. Eventually, through hard work and luck, the signal is repaired in time to cement Australia's place in the history books.

Tilting Toward the Moon

Graeme Wood's astute cinematography constantly moves to the skies, showing in one scene the Dish and the distant Moon in a single shot. Often supporting these images are sounds of radio communications between Parkes, Houston and Apollo 11. The Central West's vast plains and skies constantly feature as they are contextualised through a Dish that tilts and turns towards infinity.

During night time, the Dish is lit up like a singing Christmas tree, and its radio transmissions become a familiar soundscape in the film. In one memorable scene, the soundtrack becomes an even more complex series of acoustics as an operatic voice is superimposed over the radio murmurings. The Dish is equally grand in this passage as its night time illumination reveals its technological details. In this standout scene, the Dish turns 360 degrees like an operatic solo or a ballerina's promenade. We also witness the impressive circularity of this structure fade out only to be replaced by the full Moon. The drama of this sequence continues as we witness moving space particles edging ever closer toward the lunar target. The operatic soprano bequeaths upon the scene a tone of grandeur and majesty. Such breathtaking sounds and images prepare us for the film's climactic moment: lunar touchdown. But this is far from being the end of the film as the weather has a final say.

On the day of the Moon landing, gusty winds threaten to damage the Dish. This potential disaster is another opportunity for laconic humour as we learn that the Parkes Observatory was chosen on the basis of its temperate weather. Once more the team inside the Dish are tested as they try to navigate destructive winds after learning that NASA has lost connection with its Goldstone station, another strategic outpost tracing the trajectory of the Apollo mission. The scene is pregnant with anxiety as the three Australians and the American try to manoeuvre the Dish against gale-force winds. In preparing us for the team's success, the soundscape gradually builds toward a crescendo as the Dish's television monitor registers fleeting images of Armstrong's walk.

At this crucial moment, all those in Parkes, including the visiting Prime Minister, are glued to their televisions. The camera records many individuals and communities watching their televisions: school children, country women and families. This daring feat of science is inscribed upon their stunned faces. Actual footage of people watching televisions across the world is spliced into this film, which ultimately reinforces the global significance of the event.

The penultimate scene of *The Dish* provides a shot of the Earth bathed in a darkened space. We assume that the image is taken from the vantage point of the Moon. This reversal of perspective compounds the enormity of the feat as we are prompted to contemplate our own finite world and night skies from a lunar point of view. However, Sitch does not end on this existential note— instead, he chooses to celebrate human relationships as the binding element that enabled the achievement.

While *The Dish* chronicles Australia's technological involvement in the Moon landing, it never loses sight of the human factor that made this daring experiment possible. It is also a particularly important film for its recognition of regional Australia's role in this epoch-defining moment. And while there are some inaccuracies in the film—for instance, the first few minutes of the television broadcast originated from Honeysuckle Creek near Canberra not Parkes—it nonetheless captures the true spirit of wonder and adventure that still survives today as we gaze upon the big night skies of the Central West knowing that Antipodean space played a key role in the 1969 Moon landing.

Suzie Gibson is a Senior Lecturer in English at Charles Sturt University. She has published widely in distinguished national and international journals covering the fields of literature, film and television. Her research is informed by her knowledge of nineteenth, twentieth and twenty-first century literature, as well as continental philosophy and feminism. She has written a number of important journal articles and book chapters on significant literary figures as well as popular culture characters including Henry James, Jane Austen, James Bond and Batman, where influential philosophical concepts are explored and explained through these writers and iconic characters.

Leah Ginnivan

Transit of Venus

1. This image first came to me, returning to Albury airport from a long time away, heartbroken and hungover, looking out over the dried and carp-filled weir. I pictured my place in the Milky Way galaxy, far out on a limb. And everything from our past, present and future scattered across that galaxy, scattered but connected, held together by the forces of space-time.

2. At my saddest, age 15, I was in search of an ideology that would help navigate my experience of my places, between the bush I grew up in and the regional town where the dog food factory vented warm and meaty air into the low troposphere. I felt every day stuck in an inescapable loop and hated myself for not advancing. I hadn't yet learned how recurrence could benefit a human soul.

3. In this time, my younger sister was crazy about penguins. That would have been when I heard about Magellanic penguins, in one of the assorted penguin texts in the house in this era. Years later in Canberra at the old observatory, before it burnt, I saw the phrase *Magellanic Clouds* and thought of the adjective *magellanic* as a speckled and dusty thing. Magellan himself doesn't really deserve the penguins or the clouds—they were already known and named by others. Perhaps in the years after his death, other things were named magellanic, too, and now those names have been eroded away by social forces, leaving only esoteric galaxies and uncontested penguins bearing this name.

4. A boundary creates a frame for understanding. One of the probably two occasions we did science experiments at the tiny valley school involved measuring a square metre and contemplating everything inside it. How many beetles? How

many types of grass? What about witchetty grubs or the homes of trapdoor spiders? I found an owl pellet in my metre, a perfect owl analysis of the local biodiversity, an opalescent jewel made of insect shells. When my audit of the metre was over, it went back to just being an unbounded patch of grass in the eyes of the world while I was free to live my mobile life with a name and ambitions. I felt a strange sense of guilt—it was still mine, to me.

On a larger scale all the paddocks and dams on my family farm have names: the Marriage Tree Paddock, Kangaroo Slope, Banksia Slope, and the same thing with the dams (a new dam was built and I had the honour of naming it, choosing Heron Dam because of the white-faced heron that had her eye on it from day one). I know these places so well. The secret parts of paddocks where it gets marshy, where you wouldn't spot an incipient blackberry bush from the road, where there's an ambiguous skull of a marsupial (too small to be a kangaroo) or a strange new plant has blown in on the wind. Shady spots, places to jump between phalaris outcrops, the places where you saw, probably will again see, a snake. There's a creek strangled in applemint and weeping willows: my grandfather was staunchly anti-erosion and sceptical of native plants' capacity to hold a bank together. Around the corner, there is a ferrous cliff with a thicket of black wattles—invasive, truculent and springy—claiming back that same dirt.

5. One afternoon, I was holding him and found I needed to talk about one of my central images. I was five years old, and we were getting out of the truck in the Cypress Tree Paddock. The ewes were to be mated with the curly horned merino rams. The rams spent their days reclining under a dark thicket of cypress with branches they'd trimmed to the exact height of a sheep's nibbling range, and one of them was missing. He wouldn't come out though we could see him vaguely in the dark. I was the exact height of a sheep's nibbling range and fit under the cypress neatly and was sent in to get him out. It smelt bad. I kept walking, my dad sweatily yelling things in the pre-cypress glare. Over maybe five seconds everything came into focus. The

cold opaque tannin cypress smell, the dead smell, hot summer, a dark dark green. The ram was a skin bone bag, flies crawling in and on and through in maggot form, suspended and trapped for who knows how long by his horns on a low cypress branch. It was a biblical image of horror.

Lying across my lap he listened to my image and, as was often the case, had a counterpoint: a wood duck had tried to lead him away from her nest by doing a distraction display (also known as paratrepsis) common amongst ducks. She stuck out her wing, pretending it was broken, hobbling along. Except the wing was devoid of feathers, a skeletal fan of keratin, such that when it came to the part of the display where she was meant to fly away leaving you empty-handed, she could barely take off. A display of death in the evasion of death.

6. It's strange to me that back then I felt comfortable being in love with a different place when I felt so little about my own. Driving around his place, California, I witnessed all the songs, movies, geopolitical issues and famous lives manifest in every place, from the dystopian feedlots on the Interstate 5 to the overtly palatable and glittery scenes of Highway 1, feeling elevated and sun drunk. I complained to him that Australia wasn't like this. To his credit, he didn't accept it for a moment.

7. I tried to find other people to talk to. One man blankly refused to believe me when I told him what I'd recently learned: that all true eels (genus *Anguilla*) are catadromous, spawning and hatching in the ocean and travelling as glass eels then elvers up the freshwater rivers. Briefly, I had hope for a horticulturist who smelled just like a freshwater creek, cool and watercressy. When he dumped me I'd felt like a disintegrating plastic bag in a stormwater drain. I lay on the grass outside the old church in Abbotsford and breathed the air from the lemon-scented gums. I tried to console myself with the fact that he couldn't even be drawn into conversation about trees, not even the saw-leaves of a *Banksia serrata* we'd passed on a walk. Eventually, it all seemed like a dream.

8. Later, I came across the useful concept that place is space made meaningful (then: how do we make *things* meaningful?). A place is an ongoing process, and it does not have an inside or an outside. When I came home, I walked around a small lost altar of crumbled chimneys surrounded by figs, mulberries, and garlic and wild rose, the remains of a Chinese gold miner's lonely home. I felt my ignorance driving between Gooramadda, Yackandandah, Barnawartha, Baranduda, names I'd taken for granted my whole life. Sometime at least two thousand years ago, a local artist (perhaps of the Pangerang people) painted a thylacine, ochre on granite, in the hills behind what is currently my parents' farm. I found a stone hand axe in the bush and wondered how I had missed all this before.

9. Cultures that didn't write had other ways of remembering. One of the most widely used techniques is embedding stories, songs, and information into a landscape, real or imagined. In some cultures, the initiation begins by induction into creation myths and their evidence is all around you. Later you move through more objects and their stories; seeing them re-teaches you the story. Thus, to remember the world, you move through it: a songline.

   It is plain reality that every square metre was already Place tens of thousands of years ago. And so everything colonialism did to destroy Aboriginal people, from stealing the land and kids to purposefully eradicating languages, was also a destruction of the meaning of place and knowledge, an attempt to enforce that *terra nullius*. Where I grew up, there are places named for massacres and after murderers, and we speak these names alongside Aboriginal words without hearing the sounds in our own mouths. In other words, white official history is an organised crime, still barely comprehended by those of us growing up inside it and perpetuating it.

10. Therefore, I don't sympathise with the environmentalists who fixate on wilderness. The fantasy of being the first person to know a place is a craving to erase our real history, and to start

again with less culpability. I think of the chorus' lines in *Oedipus Rex* that I memorised in my education under the dog food cloud.

WHO WALKS HIS OWN HIGH-HANDED WAY, DISDAINING
TRUE RIGHTEOUSNESS AND HOLY ORNAMENT;
WHO FALSELY WINS, ALL SACRED THINGS PROFANING;
SHALL HE ESCAPE HIS DOOMED PRIDE'S PUNISHMENT?

(it's still always in caps in my mind)

11. Doomed pride is a satisfying aspect of Australian colonial history, and I think about it in regard to the life of Joseph Banks, an assistant to a certain megalomaniac space-seeker. Captain James Cook's first voyage was an undercover operation to find Australia, and the cover story was to observe, from Tahiti, the once every 243 years transit of Venus across the sun, so we could finally figure out how far away the Earth is from our local star in the spiral galaxy (149,597,000 kilometres). Banks, very young at the time, performed the role of learning the local languages, keeping the records, and ensuring the purported purpose of the voyage was upheld. By the time he got to Australia, Banks already had described a varietal of 'penguin', the great auk, which have all been dead for nearly two centuries now. And by the time Cook (the doomed one) was killed in Hawaii, Banks was back home, having the *Banksias* named after him, spending his days advocating for Australia's new meaning to be made via a transformation into a vast and inescapable penal colony.

12. A feature of Australian colonial art is the strange attempts at drawing native animals. Early drawings of kangaroos, for instance, resemble scoliotic Weimaraners. Early colonial artists couldn't paint the grey-blue of the ironbark forest or the eerie chartreuse of a cherry pine tract. They needed some time to erase their projections of clean lines of horses and hounds, to come to terms with the shadowy hemispheres of the marsupials and the geometrical disarray of the eucalypts. There doesn't seem to have been the same issue among the botanists drawing individual plants, and Banks' copperplates in his *Florilegium*

hold up to the eyes of born and raised Australians. *Banksian* is not a widely accepted adjective, but, if it were, it would refer to tea-coloured parchment and international scientific scholarship in a vague state of decay.

13. In the summer of the Ram, my grandmother was dying. Banned from the house, I was learning to dive and could not figure it out. Failing to bend, I smacked directly against the surface tension. But once I learned to form into an arc and then a line, I did it again and again.

    The best part of swimming in unmoving bodies of water is experiencing the temperature gradient. I loved (and still love—it bothers me how difficult it is to say something happened in the past *and is still happening*) figuring out where the sun-warmed layer meets the intermediate zone meets the cold dark part and then triangulating the perfect angle to dive to experience these layers. We did it, again and again, that summer until the behaviour was automatic. That summer, paddock appreciation was reprieve from this. Stumbling into the crisp blonde paddock, I found a nest of wild baby ducks (mother's wings intact), not too far from where my younger cousin saw a satin bowerbird's shrine to blueness a few years ago.

14. An ancient and cachectic sheepdog, Red, was dying at the same time as my grandmother. My aunt, a young doctor, was moved by his unending misery. In the orchard, under the pear tree that created (*and still creates*) a perfect shade umbrella, she gave him a lethal dose of opioids and let him die. Days later, a scratching was heard at the door. Red had slept off the morphine and was ready to have another go. After my grandmother died, one of her exotic plants finally lit up in flowers after thirty years. In the lax Catholicism of my family, we were still all keen for these miracles.

15. Another time he and I watched an asteroid shower. I was lying with my back against his chest, his arm across me. We saw the Milky Way's most visible arm, Sagittarius, spilt across the black.

I experienced an intensifying of the strange vertigo that never really leaves me, a sense that I could (*and still can*) peel from the earth and fall into the sky below me at any time. Was his mass pushing me into the sky or holding me against the grassy outcrop? We agreed that it should be treated as an empirical question.

The Ram, Aries, is not visible from the southern celestial hemisphere where I spend my days now, but I could have seen it on that day if I hadn't been falling. But conversely, the Magellanic Clouds are only visible from these skies, and on clear nights I still look for them.

Leah Ginnevan grew up in the Indigo Valley in north-eastern Victoria. She is a writer, researcher, collage artist and final-year medical student. Her favourite constellation is Orion, reinterpreted as a giant impassive cosmic face. The best night skies she's seen in Australia were while camped out at Mutawintji National Park, near Broken Hill NSW.

Miranda Gott

A star shall come

Dear Father

You gifted me with a passion and aptitude for mathematics and physics. And then you tested me in a family that expected I would spend my years tilling, planting, harvesting.

There's plenty of maths in being a farmer, my father said, and I could see his point. But the multi-factorial analysis of weather forecasts, soil moisture, repayments on the harvester and loans, and commodity pricing and futures trading, which determined his decisions about what to plant and when—it all happened in his head. His grey-capped, weather-faced head. I couldn't see the workings, only the answers.

For a time, I found a happy middle ground. A wedge-tailed eagle spooked from its prey making a liminal struggle to leave the ground, then rising on thermals in an elegant helix of $x(t)=\cos(t)$, $y(t)=\sin(t)$, $z(t)=t$. I worked the equation over and over, elaborating the maths to capture the inverse conical spirals that the bird formed.

I carried out my farm chores obediently but without interest. If you won't be farmer, join the ministry, my mother urged. She had disappointed her God by having only one child and perhaps hoped she could redeem his favour by offering me up to the church.

The night sky was my happy place. I memorised the names of constellations, planets and stars, and how they were classified. Canopus, Volans, Horologium—the names would tickle my tongue.

In Year 5, I volunteered to my class the names of the Southern Cross stars—Alpha, Beta, Gamma, Delta and Epsilon Crucis— and was beaten up after school by two classmates.

In Year 7, I compounded the error by telling the most beautiful girl in the school that I would discover a new star and name it after her. For that, I was nicknamed Yoda. "Star will I name, let me feel your Force!" the class clown would crow when I came within ten feet of the girl.

The farm was far enough out to have no light pollution, and on moonless nights the Milky Way was a streak of magic across the sky. I found the Great Rift, the dark lane in between Cygnus and Scutum, and the Milky Way's sister galaxies, Andromeda and the Magellanic Clouds. My passion for the Milky Way made me feel guilty because I knew that my lover was at her best in the long dry spells when my father became grim and the congregation offered up special prayers for rain.

We argued. God is in the soil, my father said. God is also in the stars, I replied.

My father was a practical man. I couldn't explain the awe and divine mystery I felt when exploring the night sky.

The pastor counselled me about obedience to my parents, about the lies of science, and the path of learning that ended in doubts and downfall.

I was allowed to stay on at school, but my father worked me even harder on the farm. I learned to enter a dissociative state in which my mind would wander through physics problems, chemistry equations and maths proofs. Hours later, I'd find that I had harvested a field, siloed the wheat or fixed a fence with no recollection of my actions.

My parents prayed for my soul, which I suspect meant praying for my academic failure. Their faces froze when my Year 12 results brought me a scholarship to a city university.

The years of farm work distinguished me from other university boys studying physics and mathematics. I was also brown-skinned and muscular, which made me suddenly popular with women.

I broke all the rules of my family and church that first year. I lost my virginity, my sobriety, my faith in God, and nearly lost my scholarship too.

Christmas and Easter at home were not happy. I refused to pray with my parents or attend church services, and we fought. The pastor came to the house and lectured me about obedience to my parents and to God's word.

I found paid work with the physics department during the holidays, and excuses to stay away from the farm.

My work has not been at the glamorous end of astronomy. Not searching for water that would make life possible on other planets, or producing the breathtaking pictures from the Hubble telescope, or interpreting the secretive signals from the birth of the universe. I monitored small solar system bodies—comets, asteroids and meteors—for NASA in the southern hemisphere.

One morning in late September, I received an alert for a new comet, which the automated system had detected the night before, as happened several times a year. I programmed some targeted monitoring protocols, and in a week had enough data to make some rough estimates of its orbit and size. The International Astronomical Union gave it the provisional name of C/2019R2.

The perihelion—the closest it would come to the sun—I calculated to be about 0.35 AU. Its frozen nucleus was a huge 20 x 30km, more than twice the size of Haley's Comet. This would be a Great Comet, easily visible to the naked eye, with a nucleus large enough to survive a close brush with Sol.

I had enough data by November to calculate the comet's orbital period as 2018 +/-0.012 years. But this was the eccentric and near-parabolic orbit of a visitor from the outer limit of the oort cloud, from the cosmographical boundary where bodies are only loosely loyal to the solar system. The comet didn't have high enough velocity to escape the solar system, and the pull of planets and minor planets had probably caused its orbital path to fluctuate significantly.

I spent several days programming, running and refining a model of its orbital period. At the end of the week, I had found an answer. And, with it, I had rediscovered my faith.

That night I telephoned my parents. They were surprised to hear from me, our contact of the last three years having been limited to birthdays and Christmas.

I told them about C/2019R2, leaving out the technical details. I told them they would see a comet in early December that would grow in brightness over the month. I told them where and when to look in the night sky. And, nervously, I told them that I was sure this comet's last visit almost 2020 years ago had signalled the birth of Jesus.

Stunned silence gave way to joy. I felt my father's praise and pride swelling my parched heart. My mother said amid happy tears that she was glad I had not joined the ministry. When we parted over an hour later, my last words were a plea that they keep my discovery to themselves.

I prayed for the first time in years. I applied for leave from work, planning to celebrate Christmas with my parents. I caught up on paperwork, filed reports and cleaned out the office fridge. I wrote another chapter of my stalled PhD thesis and vowed to complete it by March. I submitted an abstract for a paper about C/2019R2 to an astronomical conference in April. I phoned a woman I had met at a seminar and asked her out to dinner.

In late November, I returned to my model of C/2019R2's orbit, adding the new data compiled over the last few weeks, and for the first time, I ran it forward. I wanted to know the future of my prophetic comet.

Dear Father, this burden is heavy.

I returned to the farm the next day after driving through the night. My parents had not expected me until Christmas but were delighted, particularly when I agreed to come to the church service.

I could see that they had not kept my news to themselves.

A frisson ran through the congregation as we entered. This familiar community. Old men who had tousled my hair, and aunties who had fed me cake and dabbed mercurochrome on my cut knees. Boys and girls I had grown up with, now men with farmer's hands and women with children at hand, in arms and bellies. As one, they turned to look at me with wide grins and moist-eyed smiles as the pastor ushered us up to a front pew, warm handshakes and back slaps left and right; a kiss on the cheek from the most beautiful girl in Year 7, now 28 and more beautiful than ever.

The pastor quoted the Star Prophecy from the *Book of Numbers*: "a Star shall come out of Jacob, a Sceptre shall rise out of Israel." The comet was a sign of Christ's return, he said, and a sign of God's love for this community of good people.

I wanted to be sick, to run from the church, to curse God in a hundred black, foul ways.

The pastor called on the congregation to pray for rain, for abundant crops and fat livestock.

How could I tell them that the impact of C/2019R2 would cause a choking rain of ash and dust and a night of a thousand years that no life would survive?

Miranda Gott has been writing for ten years and was shortlisted for the SD Harvey Short Story Award in 2012. Her first play, *Kangaroo*, based on an original story, is currently in development for the stage in 2020 with the support of Create NSW. Miranda grew up in Hong Kong and the USA, has a degree in science and a doctorate in ecology and has worked in journalism, academia and policy in Melbourne, Perth, Sydney, and now Bathurst. Having moved out of Sydney eight years ago, she is still awed by the big skies of the NSW central west.

Bonita Gwyn

The Tiny Tinny Titanic

It was a dark and stormy night … No, just kidding. It was, in fact, a bright sunlit Saturday morning when my cousin came up to me with a proposal. No, not a marriage proposal, a proposal to build a boat.

Now you might ask, why would my fourteen-year-old cousin want to build a boat? (I was eleven.) Well, the answer was obvious: there was a creek not too far from our back yards. This creek and its environs (what we called The Gutters) were our playground and education centre. Us neighbourhood kids spent most of our free time there.

My cousin Raymond (for that was his name) lived with his family next door but one from my grandmother's house at the end of a very long street in our very small town. My sister and I spent every Saturday and Sunday at Nanna's place while our mother worked. We loved staying there because we had lots of freedom and there were always adventures aplenty for us to get up to. We could leave the house after breakfast and not return until we felt hungry for lunch. This pattern would be repeated in the afternoon. There was always so much stuff to do at Nanna's. Forests to explore, wastelands to survive, rivers and mountains to conquer. Whatever we thought of, we could and would do.

On this street near this creek there were many other children in desperate need of leadership, guidance and direction in how to fill their weekends. I, as the eldest of the gang—I mean group—provided this service. My cousin had not been available to perform this duty. For reasons we need not go into now, he was away for several years "at school", a euphemism if ever I heard one! He was out now, though … so back to the boat.

Naturally, Nanna's back yard became our naval dockyards while, over the next couple of weeks, the boat took shape. Captain Raymond was very good at thinking on the hop, which was just as well, as he didn't have any plans. He made them up as he went along. We had many delays too, including the usual workers' comp injuries: splinters in fingers, squashed toes from falling objects, hammer bruises on thumbs that were too slow to move, cuts from blunt saws, hurt feelings when I was yelled at, cuts from rocks hurled in retaliation, etc.

We progressed steadily throughout that cold July. And then, finally, freedom! The school holidays arrived, and we could spend even more time on our building project.

As the boat got closer and closer to completion, we kids became more and more excited. We began to plan our maiden voyage. Maybe a voyage of discovery to a new land or a pirates' raid. Anything was possible. Oh, what fun!

Finally, late on a Sunday afternoon, the boat was finished! What a glorious sight it was! The *QEII*, The *Titanic*, or any other ship you care to name, paled in comparison.

Our boat was about three metres long and a metre wide, and was constructed from timber of the four-by-two variety, tin of the flat variety, nails of the pointy variety, and tar or pitch of the hot variety to hold it all together. It was more heavy and cumbersome than sleek and aerodynamic, but it had a certain charm. The bow (front end, for you landlubbers) was square, as was the stern (rear end). To us, it was an oblong object of great beauty. Three planks of wood across the hull served as seats. The working part of the engine consisted of two rather heavy, splintery oars. And all of this was covered by a mantel of pride, with anticipation so thick we nearly choked on it.

Now I am nothing if not practical, so it fell to me to organise the crew and provisions for our maiden voyage. The provisions were easy. Some apples (because they were in the fruit bowl on the table at Nanna's), oranges (for Vitamin C to ward off scurvy), rock cakes (also because they were on the table) and a big bag of lollies (essential for energy). I included all my favourite lollies: clinkers, liquorice-all-sorts, chocolate bullets, jaffas, jubes, jelly beans and those hard flat ones with messages on them such as 'I Love You' or 'You are Cute' and other soppy things. We put the supplies in a plastic bag and stowed them in the bow.

The selection of the crew to undertake this dangerous journey was not so straightforward, however. The captain and I selected the five eldest members of our group: Raymond 14, Ronnie 11, Michael 10, Barbara 9 and myself. But EVERYONE wanted to be part of the crew. NO ONE wanted to be left behind. We were experiencing our first mutiny! Those who hadn't been selected claimed that, since they had helped to build the boat, they should get to sail in it too. Also, they threatened to tell our parents about our plans!

So, we reached a compromise. Everyone would go! The crew now consisted, therefore, of the original five, plus Patricia 7, Rudy 4 and Mary 2. But, in our haste to build, crew and provision the vessel, we had overlooked the most important thing about a boat. Its NAME! I mean it, we really overlooked it. It never got one!

After settling on the crew and provisions, the next problem was how to get our boat to the creek. When I said the creek was not too far from our back yards, well, it wasn't—if you walked or rode. But it was a very long way to haul a very heavy boat.

The captain and master builder (Raymond) came up with a rescue plan. He removed the wheels from his billy cart and attached them to a long wide plank of wood the same length as the boat. Of course, the boat was too heavy for just two sets of wheels, so he inserted another set in the middle of the plank. He then tied the boat onto this long skateboard (a man before his time!) with baling wire. Having ascertained that all was ready for the transportation and launch next day—and because it was getting dark—we went home. We were all very excited.

Early next morning (at approximately 0930 hours), the crew assembled. All of us, except the two youngest who were too small to be of use, held onto the sides of the craft to keep it balanced, and walked, pulled, hauled and carried it across The Gutters to the creek. Luckily, by the time we reached it, the crew were so hot that the cold water looked very inviting—in spite of its imaginary icebergs. We had no slip rails and the banks were wet and slippery from the early morning frost, so the launch became quite a struggle. However, with much bellowing from the captain and even more yelling from the crew, the boat slid majestically (or rather jerked, got entangled in the reeds, turned back to front, stuck in the mud and wobbled alarmingly) into the icy water. The crew hauled themselves out of the freezing, sucking, oozing mud, and climbed into the boat where they shivered like the tassels on a Go-Go dancer. (Go-Go dancing was very popular back then.)

Even though we displayed very little grace and seamanship while we were trying to get on board, we looked extremely sea-savvy once we settled onto the planks. Like old salts, really. And now we were moving! Oh, how we were moving! The boat lumbered sideways, backwards and forwards out to the middle of the creek where the current caught it, and, at last, we could begin our journey downstream towards town.

As the boat rushed headlong towards the junction of Goobang Creek and the Lachlan River, we began to plan where we'd go ashore for food to replenish the energy we'd lost in the battle to transport and launch the craft. We decided on Memorial Park because it was on the way. On the way to where, we hadn't yet decided. We nevertheless appointed a lookout to find a suitable landing spot for our lunch stop.

While the rowers rowed and the lookout looked, little Mary began to complain about being wet. Now that was unlikely because, when she had been lifted into the boat, she was as dry as a bone; and, also, she was toilet trained. Nevertheless, there was, indeed, water in the bottom of the boat where she was sitting. Maybe a few centimetres. No matter, all ships have water in their bilges, so we ignored it.

As we rounded the bend before the park, and as the lookout continued to look out for a picnic place, the pond in the bottom of the boat grew deeper and wider. The harder the rowers rowed the faster the water rose. In this contest between water and rowers it soon became apparent that the water was winning.

At this point, you would expect the crew to embark on mutiny number two, or at least have a mild panic attack. But No! Such was the faith we had in the captain and the boat (or we were just too stunned to move) everybody stayed calm and watched the water rise. We'd almost reached the park when someone, in their wisdom, grabbed the food bag. As the boat slowly went down as well as forward, we began to eat. It was only after the lookout had failed to find a suitable place to dock and the water was lapping the gunports, that the captain gave the order to abandon ship! Those who could swam to shore, and those who couldn't swim (Rudy and Mary) sat in the boat and waited.

I rescued Mary, Raymond grabbed Rudy, and we swam/dog paddled to shore. I was not, am not and never will be a strong swimmer, but I was strong enough, or lucky enough that day. When I finally struggled up the river bank I looked around to check that we had all made it safely to shore. Wow, what good luck! Not for the boat though.

We were all wet, muddy and freezing cold. Our teeth were chattering like castanets. And we had to get home in time for lunch. It was sixteen minutes past twelve. I know because that is the exact time my mother's watch stopped. I was wearing it when the boat went down. Oh, what bad luck!

By the time we'd walked back home, our clothes had dried and we were no dirtier than normal, so parents didn't suspect that anything untoward had occurred. Thank Goodness! They would have killed us if they knew what had happened.

Later that same day …

It was a dark and stormy night … No, just kidding. It was, in fact, a bright, sunlit and very cold Monday afternoon in July when my cousin came to me with a proposal. No, not a marriage proposal! A proposal to build a spaceship!

Now you might ask why my cousin would want to build a spaceship!

"I have just heard on the wireless that they landed a spaceship on the Moon," he told me. "And a bloke named Armstrong walked on it!"

Bonita Gwyn was born and raised in the small inland town of Condobolin, NSW, and still lives there. She is married with two daughters and four grandchildren who have inspired her creative writing. She began writing stories about her childhood for her grandchildren so they would know about life in a time before digital devices.

Allis Hamilton

Bereavement

In Memoriam: Justin Goldspink (1975-1992)

It was late. Everyone
was in bed. There was

no sound. No lights,
save the mighty stars above.

For days there was no moon,
only stars in the black.

I would take myself out,
lie on the soft grass,

try to fathom which
of the endless stars was him.

Bonfire

'a small smattering of stars …'
– George Mackay Brown

We huddle around winter's bright fire,
generations of a family together in the clear night.
Shadows of Cypress pines lurk like watchful ancestors.
In silence, among cousins and dogs, I watch embers drift
up and in a blink they are gone, leaving a sky-full of stars.

I did not know it then. How our lives
would become like those spiraling sparks
ignited deep in the heart of refining flames.
That, as our days unfold, we would drift
out and away into the endless stars of the world.

Allis Hamilton lives off-grid in a little hut in the Australian bush where she sings to the trees and the water and listens to the birds. She creates art, music and poetry. Her poems have appeared in *The Poetry Review, Australian Poetry Journal, Overland, Southerly, Westerly, Plumwood Mountain, Medical Journal of Australia*, and were anthologised in *Flightpath* (Hallowell Press, Australia) and in *The Creel* (Guillemot Press, UK).

Barbara Holloway

The Federation of Clouds

Before the current drought, I was living by myself out in the flat, big-sky-country of the western Riverina. One afternoon I was walking the kilometres back to the house from my mailbox on the main road. It was a hot clear summer's day, without a hint of a breeze. Head down, I was thinking hard, foot-foot, foot-foot, and gazing along the track across the open paddock. Halfway home, a shadow made me turn and look up. Above me and right to the horizon, clouds like container ships had appeared where there weren't any clouds before. They were so low I felt surrounded. They moved slowly, uniformly, as if in a herd, but their silence, and the stillness of the air at ground level disoriented me. I walked on at the pace they were moving with an uncanny feeling like I was walking on the sea-floor beneath flat-bottomed ships or immense sleeping dugongs. I reached the house and went inside. When I came out an hour later, the sky was empty. Not a single cloud, only words floating around. Paddock. Cloud, ship, grey, float, sky. Surprise.

Those flat-bottomed clouds were new to me. I belong to a long line of hunter-gatherers, farmers, pastoralists, sports enthusiasts and washing-day observers who read the clouds to know what to expect this morning, or tomorrow, or in an hour's time; an ancient tradition of water vapour and fog lovers, classification geeks, of cumulophiles whose eyes gleam at opulent storm clouds towering over the horizon, or who join groups of smokers outside buildings just to see what's happening in the skyways. (I'm not forgetting the professionals, the meteorologists.)

I like knowing how clouds enter language, participate in cosmologies, industries, beliefs and arts. Some are ways of experiencing or talking of clouds that humanity has only found recently; some are age-old. Clouds are used to name insoluble mysteries of their time and to trace seismic shifts in technology, in narratives that are at times deeply evocative, at others bare and practical. What follows is a pattern in which one aspect of clouds lights up another, beginning long ago.

The Cloud of Unknowing

For some Christian mystics, for example, the 'Cloud of Unknowing' refers to all that the mind cannot know. In that tradition, God is in the cloud.

When I was at school, with a tendency to practical mysticism, I had no problem believing in God. A question arose when I looked at art books of Old Masters' paintings that included images of God, always among clouds. How did the artists know that God chose to appear in or from behind those particular cloud formations, with those particular shafts of light? Then I went to Europe and almost every day saw those same clouds in the sky, and learned that ordinary every-day clouds were God-clouds. I was amazed, literal-minded provincial child that I was.

People also once believed that God's voice could be heard from a cloud. There was a name for this phenomenon: the *vox de nube*, or 'voice from the clouds'. Certain theologians used to claim that, although we could not see God, He could see and speak to us from behind or within the clouds. Or we might hear a voice 'crying in the wilderness' as He spoke through one of His saints or prophets. In the Riverina, I haven't heard any saints or prophets myself, although I sometimes hear a feral cat (at ground level) on cold, windy nights.

In Muslim cosmology, God has His being even higher in the sky beyond the clouds. The Prophet Muhammad first heard His voice as indescribably beautiful bells ringing down from the 'heavenly sphere'. The Angel Gabriel, God's messenger, translated those sounds into words that made sense to Muhammad so he could them write down. The whole *Qu'ran* was given that way: from God via Gabriel to Muhammad's heart and from there to his writing hand. The feeling of the original celestial music is passed on to everyone who recites the *Qu'ran* aloud following the rules and principles of correct pronunciation and stress, known as tajwid or cantillation.

From the little I know, spiritual clouds and weather clouds are closely related in Aboriginal cultures, too. I've read many stories recorded by respectful Europeans, but, at this time, I don't feel it is appropriate for me to relay them. In the aftermath of colonisation and dispossession, the issues of cultural sharing versus appropriation are still too painful, too sensitive. I am very aware, however, of the human and non-human consequences of the transformation of the pre-colonisation forests and grassy woodlands into grainlands. Including the impact on the weather, Henry Kendall, the forester and poet, observed back in 1880 that rain had decreased because, he said, there were no longer enough trees to "draw the clouds down".

Modifications of Clouds

Satellites record clouds from above. I try to imagine what that knowledge would do to the many belief-systems whose deities locate in the skies above. If you have a window seat on your next plane flight, you'll see beads of moisture crossing the triple thickness of plexiglass as the plane climbs through the clouds. The impenetrable greyness of the clouds you're moving through is very striking, as is their sun-tinged whiteness once the plane has levelled out above them. The sky seems to be an infinity of blue beyond the cloud jacket, but I can't get the beads of cloud-moisture out of my mind. I once read a report in a magazine in a dentist's waiting room about one of the driest regions of northern Chile on the edge of the Andes. Low-lying clouds pass over this place almost every night and bank up against the mountain range—and yet no rain falls. In the village of Caleta Chungungo, streams are drying up, and people are leaving for lack of water. Those who remain have set up nylon nets, mesh-covered panels the size of billboards, on the mountainside. Water from the clouds condenses on the mesh and is channelled into reservoirs. On a good night, a single panel can capture 700 litres of water. The clouds are called Chamanchacas.

Noctilucent Clouds

Another day. It's hot again. The soldier birds (Noisy Miners) in the eucalypts around the gate are making their click noises and open-throttled shrieks. Their in-y'face-manner seems hyper-righteous today. They stare at me while I'm still a hundred metres away and denounce me with their relentless shrieking no matter how often they see me. I inspect the sky anyway.

Yesterday the blue god reigned. Today the sky looks as if brushes of white oil paint were cleaned on blue Formica: nimbus and high stratus clouds, the sort that hints at rain. Noisy fat blowflies are swarming down the chimney and under the door. I climb onto the roof and tie some mesh over the top of the chimney. The mosquitoes have eased off this month, but I can't stand still outside for more than a moment before they come for me. The sky looks and feels like rain; I guess my neighbor Tony hopes it'll not fall until the last canola paddock is mown.

Tony will be mowing after dark tonight under a huge full moon. He needs to get the paddock done so the rain won't knock the seed out of the heads. He finishes as the clouds are thickening. Sirius, the Dog Star, appears between the clouds and the Moon shines through them. The term 'noctilucent' comes into my mind. It's a cloud-term that means 'shining in the night'. It doesn't apply to these clouds; it's just a great word.

Apart from the mosquitoes, there are no hindrances to sky-watching on the inland plains. I once dreamed up a plan for concentrating on skyscapes. First, buy some cheap land that can have no further damage done to it—disused infill, or an old town waste disposal site or garbage dump where there would be no distractions. I imagined watching the sky from the doorway of the site supervisor's hut on a flat plain of dust or mud with ears of buried plastic bags sticking up through the earth. I'd be able to watch the whole bowl of sky from such a place. I got less keen after I found out the smell of garbage doesn't go away for decades, and that such dumps give off so much methane gas it can be syphoned off and bottled.

I forgot how the sky is abused, too. How it's sawn into sections by jets that generate their own clouds. I've read that, in the right air conditions, it takes only twelve jet trails to cover the sky with vapour. Clouds of pollution pre-date jetliners, though. The first record I've seen was by John Ruskin, that righteous old nineteenth-century British art and sky-lover, who described black clouds he'd never seen before being blown across the sky from the vast potteries in northern England.

Satellite sensing of a cloudy atmosphere

Ruskin also wrote about the age-old pastime of seeing shapes in the clouds, a kind of theatre in the sky. In Britain, generations of weather-watchers identified clouds by traditional names: Mare's Tail and Grey Mare's Tail (also called 'Goat's Hair'), Noah's Ark, Salmon Cloud (long illuminated stripes appearing at dawn or sunset), Wane-Cloud, Wool-bag Cloud (convex or conical heaps increasing upwards from a horizontal bottom), White Tempest and Water Wagons. Scattered patches or streaks were called Prophet Clouds. I haven't personally heard those names used in Australia, apart from Mare's Tails.

The British landscape painter John Constable used new names for them as he painted clouds with scrupulous accuracy that reflected a revolutionary scientific approach to weather-watching. He 'read' and named clouds in the same systematic way meteorologists use today. This approach was developed by Luke Howard, an English amateur meteorologist, who drew up a system for classifying clouds "for the benefit of Agriculture and Navigation". A sky-watcher from his teens, Howard devoted himself to "the observation of atmospheric phenomena, the forms of clouds and the laws governing their mutation," as one historian puts it. He took these observations from a fixed point in his garden twice a day, drew the clouds he saw, and made notes about them and about the winds, or lack of them. A true cloud-obsessive.

After years of gazing, sketching and note-taking, Howard pulled his thoughts together in his "Essay on the Modifications of Clouds" and read it to friends. The essay was published in 1802, with three charts of cloud formations, in Tilloch's *Philosophical Magazine*. After stating that clouds took one of three forms—sheet, fibre or heap—he introduced the names we still know them by —nimbus, cirrus, stratus, cumulus, and so on. He called the formation and changing shapes of clouds 'nubilation'—a term that didn't catch on—and showed it to be the result of atmospheric processes. In today's terms, Howard's essay could be said to have gone viral.

His system has since been adopted across the world. Even Goethe, the most eminent European writer and scientist of his time, was so impressed that he immediately wrote to a colleague in London asking him:

> To obtain for me if possible something however simple
> and scanty about the course and manner of Howard's life,
> so that I might know how such a mind had been trained,
> what opportunity and circumstances had led him to look
> at Nature with a natural eye, to devote himself to her, to
> discover her laws and to reimpose them upon her.

Goethe was also inspired to write a poem about each newly-named cloud and an elegy to Howard himself. (I can't help noticing the sexism implicit in describing writing of 'nature' as female, a 'she' who essentially has to accept whatever 'man' wants to do. But that's just how it was then so moving right along …)

To answer Goethe's question in brief, Howard was quite a guy. Born to a devout and wealthy Quaker family, he studied pharmacy, which involved a long and miserable apprenticeship, the time he began to record the clouds he could see from a single point in the garden at least once a day. In mid-life he owned a pharmacy shop in London and was also active in the anti-slavery movement and other welfare causes while writing several books about meteorology. He was also a successful developer of industrial chemicals and pharmaceuticals, which of course he'd get fewer ticks for today. In the twentieth century, UNESCO declared his "Essay on the Modifications of Clouds", like his later book, *Seven Lectures on Meteorology*, to be "part of the knowledge base of civilization". He wasn't alone in devising a science for clouds, but the success of his system was its use of the Linnaean classification system, making it readily adaptable in other countries and cultures.

Engineering a Date on the Cloud

Everyone talks about clouds, but these days they aren't usually talking about water vapour. They mean 'The Cloud'. This 'other' cloud, even more intangible and inaccessible for us folk on the ground, has been with us since the 1990s, running on the Internet, and keeping our data stored. Not visible, not floating above the earth, its workings are most often inside vast concrete structures built into hillsides or across industrial complexes. It's cloud-like only in that it continuously changes as people change their data and computing activities on their various devices.

Our digital and physical worlds are getting more and more intertwined. This is the biggest change in cloud-talk since Howard. Experts speak now about "Complex Systems and Clouds: Middleware for Edge Clouds and Cloudlets", for example, or in Wikipedia's language of Community Cloud, Distributed Cloud, Multicloud and Cloud Engineering. These are part of our reality, too.

In 2005, British cloud-spotter, Gavin Pretor-Pinney, founded the Cloud Appreciation Society to unite cloud-lovers around the world. The Society's members "fight blue-sky thinking and celebrate the accessibility of cloud-watching to everyone blessed with vision". Its website is learned, literary and full of photos. I haven't joined, but I understand its approach.

One fine day last spring, a visitor arrived at the farm and we walked around the wheat paddock, heading north, chatting. Unnatural stillness stopped us, then about a hundred crows suddenly streamed into just one tree nearby. The low ridge turned a strange sulphur-yellow. The wind hit—which certainly ended silence—and a low black cloud spread across the ground as if someone were pulling a vast scarf over us. The all-encompassing 'scarf' was followed by a wall of rain. Like Virgil put it centuries ago, "down headlong falls the sky in sheets".

We slogged home, hardly speaking. The storm didn't last long and, had we looked at the Bureau of Meteorology's website, we might have taken a jacket. But somehow not having seen it coming was better. We've had little rain since then.

My celebration of the sky is tempered now by a fear of what else might be coming. Climate Change fills our sky-watching with anxiety: those storm clouds are blacker and thicker than any we usually get around here, aren't they? Is that thin cloud-cover or dust across the horizon? Is that band of cloud wide enough to bring decent rain? And—what will our future be like?

References

Anon. 'Chamanchacas.' *The Economist*, Feb 5-11, 2000, p. 83.

Badt, Kurt. *John Constable's Clouds*. Trans. Stanley Godman. London: Routlege and Kegan Paul, 1950.

Berndt, Ronald M. and Catherine H. Berndt. *The Speaking Land: Myth and Story in Aboriginal Australia*. Ringwood, Vic: Penguin Books, 1988.

Paul J. Crutzen, ed. *Clouds, Chemistry and Climate*. Berlin: Series NATO ASIS, 1996.

Dawson, James. *Australian Aborigines: The Languages and Customs of Several Tribes of Aborigines in the Western District of Victoria, Australia. Melbourne, Sydney and Adelaide: George Robertson*. 1881. Facsimile edition, Canberra: Australian Institute of Aboriginal Studies, 1981.

Day, David. *The Weather Watchers: 100 Years of the Bureau of Meteorology*. Carlton: Melbourne University Press, 2007.

Gergis, Joelle. *Sunburnt Country: the History and future of Climate Change in Australia*. Carlton: Melbourne University Press, 2018.

Stephen Levy. 'Bill and Andy's Excellent Adventure II," *Wired*, April 1994.

Virgil, Eclogues. *Trans. V. Gordon Childe*. Harmondworth: Penguin Classics, 1978.

Minnis, Patrick et al. *Cloud Properties Derived from GOES–7 for Spring 1994 intense observing period using version 1.0.0 of ARM satellite data analysis program.* New York: ACM Digital Library Association for Computing Machinery, 1995, https://dl.acm.org/citation.cfm?id=886408, Last accessed 25 September 2019.

Barbara Holloway most often writes creative nonfiction focused on the natural world. Her previous work has appeared in essay collections and in journals such as *Westerly, Southerly, Fusion* and *AJE*. She's a Visiting Fellow at the Australian National University and divides her time between Canberra and the bush block that's part of *BREW; Bush Residencies for Ecological Writers*. "The Federation of Clouds" is part of a longer manuscript.

Natalie Holmes

Fallen Star

Beneath our radiant Southern Cross,
We'll toil with hearts and hands;
To make this Commonwealth of ours
Renowned of all the lands …
– from Advance Australia Fair

Every week, at least one woman is killed in Australia through domestic violence by a partner or former partner. Yep, it's some lucky country, I think, as I look up at the night sky, which, once-upon-a-time, held so much promise for me. When I was young, I would marvel at our galaxy and wish upon falling stars. I'd gaze at the planets and wonder about the universe and all that it contains. I tried to pick out the different constellations, but only one was ever prominent for me: the Southern Cross. It's still there, and I'm still here, but too often my eyes have filled with tears, and my view of the stars has blurred. These days, I wonder how I can reach them to escape from a world that brings me so much pain and suffering.

In London, I couldn't see many stars because of the smog and city lights. That's where I met my 'Prince Charming'. He swept me off my feet. At the start, it was perfect, like all fine romances. We'd walk hand in hand beneath the black English skies. He was my hero, my Rob Roy. I wanted his babies. But I also dreamt of my homeland, so did he, and soon we were back beneath the Southern Cross. We married quickly, Rob and Rosie, happily ever after. The End?

It wasn't too long before the fighting started, though. I'd smooth it over, say Sorry, and all would be forgiven. Until the next time. What started as an adventure of two kindred souls quickly became a nightmare. He didn't go to work, he spent all my money, he crushed my spirit, and, most nights, he drank himself into a coma. My friendships died, my work suffered—and my body bore the bruises.

"Rosie, where are you, you stupid cow," he roars. "Wait until I get my hands on you. You are going to pay, you fat slag!"

Every night is much the same. There are some variations, of course. Sometimes, he throws the dinner I cook for him at the wall; other times he smashes my face into the floor.

More often than not these days, I run off. Sometimes, I walk the streets with our dog, Franky, until my legs ache as much as my heart. Or I hide behind our back shed and gaze up at the stars. I generally return home only when it's safe to do so—when he has passed out. Often I sleep in the spare room, but in the middle of the night he'll stumble in to find me, to reclaim his 'possession', then ramble on incoherently. Some nights, he'll vomit (or worse), prompting me to move once again. An endless round of drunken musical beds with me neither drunk nor enjoying the party game.

Tonight, my impromptu stargazing has been brought on by his crazed ranting and raving. I've found refuge again in the overgrown garden behind the back shed, where I now crouch amid bald tyres. The long grass makes my face itch. Franky is frightened, too. He sits quietly beside me in the moonlight, licks my hand, and comforts me with his warmth and kindness. He is my best friend. I shiver as I hear the back door slam, and my name being shouted once again.

"Rosie!! I will find you! You better answer me or else!!"

I nervously stroke Franky's head for comfort, and hug his body close to me. My bladder aches but I can't risk going to the toilet, so I pull my shorts down to relieve myself in the dark. I'm also a little peckish, as I've not had a chance to eat anything this evening. I want to go into the shed and read by torchlight. Books take me to another place and help me dream about a better life with a loving husband. I fantasise about converting the shed into a private flat. Even with its cobwebs and dust, it's nicer than my home where I'm choked, punched and punished every day.

As I stare at the stars, I wonder how this came to be. I'm from a nice family, I have a good education, I just don't understand why I married a monster.

No one suspects my shameful secret in our small inland town, though. In public, he is as charming as he was when I first met him. In private, he thrives on my misfortune. A true Jekyll and Hyde.

I've thought about telling someone. But who? My parents wouldn't understand. Their own marriage was happy. I have friends who might have helped, but I'm ashamed to say anything. I know there are professional services available, but I'm not sure where to begin. I feel truly stuck. Where would I go, anyway? And what about Franky?

But then I look up at the sky again and start thinking about all the brave things I've done in my life. The stars take me back to my childhood, and, even though I feel uncomfortable lying in the grass, my dog by my side, I begin to feel sleepy. The house has gone quiet. Rob has probably passed out.

Then I hear the back door creak open, the sound of him stumbling into the back yard. And Franky barks and reveals our hiding place. I can see my dark knight approaching, a forbidding shadow against the starry night. I'm shaking. Is it too late to wish upon a falling star?

Natalie Holmes has worked as an editor, writer, journalist, literary judge and public relations specialist in Australia and overseas for more than two decades. She now runs a successful communications consultancy, nh publishing, in Dubbo while wrangling an active preschooler. She loves watching the news, correcting people's grammar and eating chocolate from her secret stash. This is her first anthologised work.

Katie Hopkins

All That We Are

The child came into the world facing the sky. As her eyes blinked open, she saw first the great expanse of darkness above her, scattered through with twinkling diamonds. Then she saw her mother's face, took her first breath and began wailing. She had been so eager to be born that she could not wait for her mother to reach the hospital. The car engine groaned on the side of the narrow dirt road, where her father had pulled over in such a hurry. The Moon and the stars smiled down on this little scene, so far below them. This child. This life.

What did the child think when she saw those twinkling spots above her? Did she wonder, even at that early age, what they were? In time, she would forget that picture—no one remembers the first things they see—but she would always experience the same comforting feeling whenever she looked up at that world above her. The world filled with countless other worlds that she first saw from a hill somewhere in inland Australia.

She was running up the hill far ahead of her parents on the rocky ground. Too far ahead. She was only five, and they loved to see her so happy. They let her run further and further ahead of them, while they carefully kept an eye on her. The sky was filled with clouds, blocking the light of the rising stars and moon, and darkness covered the ground like moss. She couldn't see very far in front of her, but running like this set her heart brimming with joy. The fresh, cold air filled her lungs and her heart took flight, soaring like a bird through the trees surrounding her and up, up, up into the sky.

The family had been visiting the place where Astra was born, as they did every year on her birthday, and had wandered further and further into the bush. The darkness ambushed them as they walked down the rocky hill to the car.

A sudden change in the slope took the child by surprise. A quick gasp filled her lungs with cold air like ice. In a heartbeat she had lost her balance. A loose stone set her tumbling down the hill— right into a tree. Her parents started running as soon as they lost sight of her, cursing themselves for letting her get so far ahead

"Astra?"

"Astra! Are you alright?"

"Where are you?"

"Astra!"

They had named her after the stars she was born under, and it was those same stars that led them to her. They saw her when they reached the bottom of the hill. She was sprawled on her back next to a tree, staring up at the sky. They ran to her and knelt down beside her.

"Where does it hurt?"

"The stars are so pretty tonight," Astra replied. "And the moon is so bright. Why is it so bright only sometimes?"

She felt no pain, she said. Why would she be in pain when she could see the stars above her? But never again would she run carelessly through the bush. And nor would she ever walk again.

Astra watched the stars slowly moving across the sky. The other children ran, yelling and fighting, for space near the campfire to toast their hard-won marshmallows. She wheeled herself further back from the light and the noise, into the shadows at the edge of this happy scene. Here she could watch the deep blue above her in peace, without the chaos within the circle of firelight. She looked up at the glittering stars. They calmed the hectic whirlpool of her thoughts and she momentarily forgot her sadness, her embarrassment. Her mind flicked through the events of the week-long bush-camp, and how no one had wanted to be her partner for the bunk-beds.

"At least that means you're guaranteed a top bunk!" the camp instructor had enthusiastically told her, before flushing with embarrassment at the insensitivity of the comment—how could a disabled child climb the ladder to the top bunk?

The staff had promised that "adjustments would be made for your child in the camp activities [because] we most certainly don't want her feeling left out."

Now, on the last night of the camp, she was yet to see any adjustments.

For the entire week, she had watched the other students enjoying giant swings, high-ropes and muddy adventures while she stayed on the dry ground, with the "all-important" job of taking photos for the school newsletter. She had spent the whole week watching her fellow students laughing, yelling and playing—as twelve-year-olds will—while she sat alone, embarrassed by their sympathy and hurt by their awkward attempts at helping. She was officially part of this camp, but not really part of anything.

That evening, her teacher Mr Pearce approached her with a skip in his step and a sympathetic smile on his smooth face. "I've noticed you watching everything very carefully at this camp, Astra."

She tore her eyes from the stars and silently nodded.

"Why don't you write an article about it for the newsletter? You could choose which photos to include, too. You've taken most of them anyway."

She took a long look at the sky above her and made up her mind. "Alright," she said.

Mr Pearce had shown her a passion for writing that would push her through a degree in journalism and take her far away from home.

When the offer had come through, she wasn't sure what to do. A year was a long time to be so far away from everyone she knew. But she reasoned that the sky was the same in the crowded city of Bengaluru as in the Australian outback, and she had nothing holding her to the soil of her home. So she accepted.

The government has agreed to reconvene next week. Finishing her article, she emailed it to her editor for publication the following morning. She leaned back in her chair, sighing with relief and exhaustion after a long week of interviews and government meetings. Wheeling herself outside, she looked up at the stars. A beautiful, clear night, with a cool wind blowing in from the East. She inhaled deeply. The stars reminded her that Bengaluru, or Bangalore, with its busy streets and Indian culture, was not so far away from Australia but, after being stationed in this city for a year, she now missed home.

She wheeled around the side of the house and up the driveway, passing the front steps, which had caused her so much anxiety for the first few months until she had been able to procure a ramp, and then rolled down the path, with no particular destination in mind. The exotic smell of spices filled the air, and many times she had to dodge careless pedestrians or cars. Ten million people going about their lives created a permanent buzz in the city air. She found herself in a park, almost empty in the night and a far cry from the bustling mayhem of the day. The buildings lit up the city, brighter than she thought possible, like a thousand stars clumped together. Tonight, however, the lights seemed dimmer and the crowds quieter amongst the trees. She looked up at the stars, so often obscured by light pollution, clouds or smoke, and thought nostalgically of home. Lost in this reverie, she hardly heard him approach.

"It's clearer than it has been in a long time," he said. "I haven't seen them this bright in months. Years, even."

She turned, startled, to see a young man, about her age, with a kind smile, eyes that laughed, and that familiar Aussie twang to his speech.

"Oh, sorry for scaring you. I come here at night to watch the stars as well. There haven't been many for me to watch recently, though. The lights …" He gestured to the buildings that seemed so far away.

"I'm Astra. I'm surprised I haven't seen you before."

"Nice to meet you, Astra. I'm Sam." He held out his hand in a comically formal fashion.

"I usually come later, once I've finished my work," he explained, after she shook his hand, "I'm a journalist—but tonight is just so beautiful I couldn't stay inside any longer!"

"I'm also a journalist. What are you reporting on?"

They stayed in the park chatting until well past midnight as if they were old friends. A few nights later, she found herself taking that same path to the park. They spent another night talking and laughing.

It was in that same park, under the same stars, that they had their first kiss. It was in that same park, under the same stars, that he proposed.

She held the deceptively simple certificate in hand. Shining diamonds of despair rolled down her face, reflecting the light of the stars above. Here, in their rural home, there was no traffic. No light pollution. No bustle and noise of crowds. No spices filling the heavy air, or Hindi words scrawled across buildings and posters. Here she was alone. A sense of abandonment descended upon her. First, her father had gone when she was fourteen. Cancer, the doctors said. Then, five years ago, her mother left. They said the crash was non-fatal but they were wrong. Now her husband was gone. They shouldn't have sent him to Syria. It wasn't safe, especially for a journalist.

She turned to the darkened window, behind which their child slept soundly, too young to understand what Astra meant when she said "Daddy's gone away. He's with the angels now."

She looked at the certificate again, to the words underneath the imposing title: *Name of Deceased*. She could not look at the next words, the name. Samson. Her Sam.

Instead, she looked at the stars. Immediately, she felt the calm that always descended on her when she saw these shining miracles. Taking a deep breath, she turned her mind away from Earth, and into the sky. She sat there for a long time. Only when she saw Orion going to rest did she bring her mind back to Earth. Back to her family. She thought of the young girl upstairs who didn't understand why her father was never coming back. She felt her now-large belly and thought of this unborn child who would never know its father. She looked at the stars again. They were there for her, even if no one else was. A constant in her life that she could rely on.

Goodbye, Samson.

She wheeled herself inside to prepare for the future. She would make it the best future she could. For their children. For him.

The car bumped its way through the potholes and puddles of this rarely used road as if on its own accord. Her weak heart beat with the same excitement she felt whenever she visited this place. The place of her beginning. Her wrinkled hand brushed a strand of white hair from her face as she looked over at her daughter who was driving so placidly despite the adverse conditions. A little glass cat hung from the rear-view mirror, clinking as its beads bumped against each other, the only sound in the silent car.

Astra's window was half down, allowing in the brisk night air with its distinctive smell of eucalyptus and cows. Dry dirt blew up behind them, covering the road in a brown fog. This drive was one she knew well. It filled her with peace and happiness and a sense of belonging. This was her place in the world.

They finally arrived at the hill. Her daughter forbade her from carrying too much as they unloaded the picnic blankets, and climbed the gentle slope to where they could best see the night. There they lay on the picnic blankets for hours and watched the constellations move across the sky. The Moon was a ghostly crescent. Tomorrow, it would be new.

They talked quietly about mundane, ordinary things. How the grandchildren were going in school, work, the family barbeque next week to celebrate Astra's 80th birthday. Eventually, their voices drifted away and silence descended upon the hill.

In the magic of the early morning, her daughter stood up, stretched, and walked over to where Astra was sleeping. It was only after she couldn't wake her, only after she checked her pulse, that she realised her mother had returned to the stars.

Katie Hopkins is a senior high school student in Western Sydney. This is her first publication, though for the past three years she has taken part in the Write a Book in a Day Competition to raise money for kids with cancer. She enjoys writing in her spare time and is inspired by the everyday interactions that she sees around her. She aims to study law and hopes to continue to publish and to inspire others in the future.

Shannon Jade

SkyWorks

The air hung still over the vast inland, and only an occasional earth-bound light penetrated the darkness. Hundreds, thousands, millions of stars were sewn into the night—until, suddenly, their threads came loose. All of them, and the Moon too, came racing towards the Earth.

The river lapped at Eddie's ankles. His neck was tilted upwards, his face angled towards the sky, and his eyes were closed. Although he didn't see anything peculiar, he felt the universe shift. He had always believed that it was easier to be seen where fewer people were looking. Out here in the bush, you had to care, and you had to matter. You couldn't be just another droplet in a sea of anonymity. People in the cities didn't notice the falling stars right away. Too much glitz and glamour and too many street lamps had hollowed out their skies so that the emptiness of this night did not seem unusual. People in the bush recognised the change the moment it struck.

Eddie's hand reached skywards. "Look, Ayla. The Seven Sisters—they're putting on a show for you."

His gaze flitted between the sky and the twinkling eyes of his tiny daughter sitting on a log beside the river. She kicked her little legs back and forth with delight.

"Why are the Sisters in the sky, Papa?"

"Every night, they're chased into the sky and run all the way across it. Look how beautifully they move." Again, Eddie's powerful hand swept through the night air, following the arc of the sisters' nightly dash.

"I want to go to the sky, Papa. I want to dance in the sky." Ayla jumped down from the log, spread her arms out wide and twirled joyfully, reaching out to hug the night.

Eddie turned his back on the sky to watch her. "You, my dear, are already a star," he told her. And at that moment the real stars began their descent.

"Papa!" Ayla cried. Her small hand grabbed frantically at the air but did not succeed in stopping the fall. "Gone!" she said. "Gone!"

The night offers gifts for those who stay awake to receive them. In rural New South Wales, under a newly starless sky, there was no such thing as a crowd. On this night, there was only a captive audience of two.

"Have the Seven Sisters lost their balance, Papa? Have they accidentally pulled down all the light?"

"All I know is this," Eddie said. "Nothing falls without landing somewhere."

The stars had made no crash, no clunk, no quiver of a sound, but Eddie and Ayla were convinced that they must have struck the ground somewhere, and it was their job to find them. They equipped themselves with nets, jars and Tupperware containers to hold the stars, and bundled up old pieces of cloth and their favourite t-shirts to grab them. Neither of them knew how you were supposed to catch a falling star or carry it, nor what you were supposed to do with a star once you'd captured it. Such practicalities didn't occur to them—because, sometimes the How, What and Why didn't matter. The answers would arrive by action.

"Wish, Papa! Wish!"

Fathers don't usually take orders from their children, least of all from those who were wandering around outside long past anybody's reasonable bedtime. But this was not a usual night. When Ayla commanded her father to shut his eyes and wish with all of his might, he didn't quibble. He scrunched up his face and sent his dreams out into the air. By the time he opened his eyes, his daughter had found the first star.

Stars float on water, it seems. Where once this star's reflection had faithfully shone, the real star now bobbed gently up and down on the river's surface, glittering as it struck each tiny current. Ayla waded into the river fully clothed and scooped the star into a jar.

She returned to shore and handed it to her father, on whose wish this star had already been spent.

Sometimes, on very rare and inexplicable occasions, when something truly incredible has cracked through the monotony of life, an internal map spontaneously appears. Not a directional guide that anyone can read, nor an instructional manual that people turn upside down and draw squiggly lines on as they attempt to make sense of it, but a kind of map that simply lives and doesn't need to be understood. Eddie and Ayla did not need a physical map on the night the stars fell. They did not stick to routes they knew well. Instead, they followed their feet, their guts, and the kind of directions they didn't need to question because they weren't designed to be understood.

Eddie's voice cut through a not-quite silence: "Ayla, look up!"

A small bird twittered over her nest with all the pride of an excited new mother. Her eggs shone brightly, iridescently. They were lit by something startling, something that the mother bird, so cheery and pleased with herself, was claiming as her own creation.

"Wow!" Ayla's face glowed as she turned to Eddie.

He was already lifting his net to capture the star.

Ayla placed a small hand in his arm. 'No!' she said.

"What's wrong?"

"It's a nest, Papa. You can't disturb a bird's nest. Look at the mother birdie. She thinks it's one of her babies."

"Can't leave it there either, Ayla. Those eggs look fresh. The heat coming off the star will either cook them or hatch them before the night's through. I promise to be gentle."

Extracting his arm from Ayla's grip, he slowly leaned forward and prodded the nest. The mother bird flared her wings protectively to shield her eggs and the star.

Ayla pursed her lips and whistled. The mother bird tilted her head to one side as if to say 'I'm listening.' Ayla whistled again.

Slowly, almost painfully, Ayla the Whistling Wonder lured the mother bird from her nest. As soon as the bird had stepped just far enough from her eggs, Eddie jabbed the net forwards and captured the dazzling star. He and Ayla leaned over it and wished with all their might.

Eddie's hands were covered in calluses and blisters. They were used to hard work and heat. His feet were used to rough terrain and hours of steady walking. His eyes were used to looking for things that others neglected to see. If he was supposed to find another fallen star—and he knew that he was—then he trusted that he would. He found it resting on a bare patch of ground. He bent down to examine it. Its shape was unusual. Its light burst out not in a smooth sphere but in flickering movements, and in the clear figure of a girl. Not worrying about being burned, he picked up the star in his bare hands and showed it to Ayla.

"Her arms," he whispered. "Her legs."

"Her shining face," Ayla said. "Is it one of the Seven Sisters?"

Eddie and Ayla held that star tightly and shared a wish that the Sisters could return to their home in the sky and take all the other fallen stars with them to repopulate the night.

They collected seven stars, one by one and piled them together. Collectively, they formed a mass of light and energy that radiated warmth and power from a united central core. When the very last star had been placed at the top of the pile, and the very last wish had been made, the mass began to shake. Slowly, the bundle of stars lifted off of the ground and began to separate. One, two, three, four, five, six, seven. The Sisters picked up speed and leaped skywards as streaks of light. Their glowing trail splintered into eight, nine, ten, twenty, fifty, one hundred, two hundred, a thousand, a million, billions and trillions of stars.

Almost as seamlessly as they had fallen, the stars soared back home and restitched themselves into the night. Here on Earth, they left wishes. They left truths. A new story for a new generation of stargazers.

Shannon Jade is a young writer based in Atwell, Western Australia, who believes in the real-world magic of storytelling. Most often found with a book in one hand and a cup of tea in the other, Shannon hopes to share her stories with the world and to create new worlds with her stories.

Neville Jennings

Stargazing at Bogan Gate

On ABC radio one day I heard a songwriter tell a bizarre story about a scientist—let's call him Norman—who worked at the Parkes Radio Telescope. Norman was a conscientious worker, but he seemed to have little social life. So his colleagues organised a blind date for him at a Chinese Restaurant in Parkes. Because they weren't able to find another astronomer to dine with him, they invited Debbie the Astrologer from the Crystal Healing Shop to be his blind date. When the two met, they quickly established that they viewed the night sky from very different perspectives. At the end of the meal, Norman handed over his business card and watched longingly as Debbie sauntered down the main street. Over the next few weeks, his colleagues noticed that he was more than usually quiet and eventually deduced that Debbie had not called him. They went to the Crystal Healing Shop to find out what had happened. "Look, he was a really nice guy," she told them. "But I couldn't possibly have a relationship with a Sagittarian."

This strange story caused me to think back to my own experience of the Parkes Shire. In late January 1961, at the ripe old age of nineteen, I took up my first teaching appointment at Bogan Gate Public School some forty kilometres west of Parkes. I arrived at Bogan Gate by train with a hot little telegram in my hand and met the stationmaster's wife, who also happened to be the secretary of the local Parents & Citizens Association. My telegram stated that I could get accommodation at the local hotel but, as my gaze shifted from the large wheat silo to the village, it was obvious that there was no hotel. As the lady from the P&C explained, the pub had burnt down over the school holidays. But not to worry, she told me, I could board with the McPherson family, who owned a farm out of town and had children at the school. It was during my first nights on the farm that I saw the night sky in all its central western glory.

I'd grown up in Sydney where, even by the late 1950s, smog blotted out many of the stars. Despite the haze, however, we'd all spilt out into the street to see the first Sputniks pass overhead in the late 1950s. We became even more interested in the night sky when the Americans entered what they called the Space Race.

At Teachers' College in Wagga Wagga I had taken an elective in astronomy and particularly enjoyed the classes in spheroidal geometry, which enabled us to track the constellations. I was keen to impart some of that knowledge in my teaching. At Bogan Gate, I taught a succession of combined Year 4/Year 5 classes and, as part of the science curriculum, was able to include a little astronomy. I recall one attempt to explain the solar system and eclipses by blacking out the classroom and using balls of various sizes, including an orange and a ping pong ball, to represent the planets and our moon. The sun was a light bulb. In those days, the solar system included the planet Pluto, of course.

Many of the students took a lively interest in the Space Race at this time. In one of our school news sessions, on a cold winter's day, a lad reported that the Japanese had put a man into space. "There's a nip in the air this morning!" he said, with a grin. We then launched into a serious discussion about inappropriate language!

The radio telescope at Parkes was being built while I was at Bogan Gate so I took students to see it. Their initial response was disappointment because they expected to be able to see the stars close-up. This led to a lesson on radio signals.

I drove a red Standard Ten car with a canvas hood at this time and made weekly visits to Parkes to play hockey, purchase LP records and enjoy a slightly more varied cuisine. The corrugated dirt road between Bogan Gate and Parkes was not kind to the car, though. I was soon appointed sports master at the school and helped the local policeman, Tim Tyler, start a boys' club. I was also the local Cub master, which gave me an opportunity to show the boys how to navigate by the stars, the Southern Cross in particular.

I was studying Geography in off-campus mode through New England University by then and had developed an interest in aerial photography and remote sensing. As NASA's space program got underway, we were better able to study the earth below using photography and other sources. President John Kennedy had launched the American Space Program in 1961, and my policeman friend had tape-recorded many of his inspiring speeches. In November 1963, when I was about to leave the district, I walked into the Bogan Gate Post Office one Saturday morning and saw tears in the postmaster's eyes. President Kennedy had been assassinated. Our world had changed, but our explorations of space continued. In 1969, Kennedy's vision was fulfilled when two men walked on the Moon and returned to Earth safely. The Parkes Radio Telescope was famously involved in transmitting images of that event. I was teaching Geography and using NASA's photographs of the natural world to highlight the uniqueness of Planet Earth.

It is perhaps fitting that I ended my career at an institution called Southern Cross University educating a new generation of teachers. Now in retirement, I reflect on the decades I've spent with school and university students, and, whenever I gaze at the bristling night sky of rural New South Wales, my thoughts often turn to that little school at Bogan Gate, where I began my teaching career.

Neville Jennings is a retired educator living in Murwillumbah, northern NSW. He began his career as a primary teacher at Bogan Gate Public School in 1961 and later moved to Merriwa Central School. In the 1970s and 1980s he taught Geography and other Social Science subjects at secondary schools in NSW, Victoria and Canada. From 1990 till 2008 he was actively involved in teacher training and teachers' professional development through Southern Cross University. He currently volunteers as a Primary Ethics teacher at Chillingham Public School.

Jane Fenton Keane

Spica

I sit here casually observing the universe
with all its cataclysms and black metaphors.
Asteroids, comets, planets and stars
share that infinite space
which remains aloof to our perceptions.
I fantasize about Scotty beaming me up,
while astronomers analyse, organize and digitise
the night and its chaos. I simply remain mesmerised,
riding a magic carpet which separates me from
Earth and higher Earth. A tapestry
woven from threads of dreams,
each line folded in on itself
like a crumpled love letter –
the sky's full of their ghosts.

I amuse myself by counting satellites
as they rotate their lenses to photograph
me watching them. Is it possible to float
to the top in the fattening orbit of space junk,
defying the laws of physics, the latest
in a series of domesticated animals
bred to feed – a cross between a robot
and takeaway food container.

The Superperfect Number Sonnet

Splashed in blood, his words are full of the long vowels of bleeding.
Tomorrow, 64 days away, all distance at once, this is how far it was.
Her eyes pure lymphocytes; whiteness beyond a blood's uncertain
call. 64 a number that cannot contain its own digits, a self-number.
Yet here we are and there we were so long ago. Between 6 and 4, 2.
6-4=2. These numbers drive me mad. Just the two of us and our four
kidneys, our two hearts (beat as one), each with their four chambers
their two halves. Each with their four ventricles and four atriums.
See how our conversations inevitably bleed. Above, where we turn to
pray in moments like this, in the Black Eye Galaxy of Coma Berenices
a spray of stars scatters in a portrait of Queen Berenice's sacrificed
locks of long blonde hair. Beneath your greying tufts, that "will you still
need me will you still feed me" question we danced to when
The Beatles were famous, dissipates beneath the urgency of here
and now, you and I, and that third party whose strong pulse must die
before your birthday in 64 days before that star in your chest blackens.

Jayne Fenton Keane is a poet whose work has been extensively published in print, radio, digital, performance, sound and visual mediums. Her published poetry collections include *Torn*, *Ophelia's Codpiece* and *The Transparent Lung*. Jayne's practise reflects an ongoing interest in exploring and experimenting with language. She was shortlisted for the Griffith University Medal for her doctorate in poetics.

Gai Lander

Cloud Moods

I gaze in the distance
Clouds on clouds

Fat, fluffy puffs
Long level lines
Fuzzy splotches
Gathering globs
Wisps wafting
Shifting spirals

Constantly changing
Moods moving over mountains
– Lucille Kraimer

I've known clouds in the skies of inland NSW all my life. When I was a child, our family lived in Armidale then Bathurst, Walgett, Narrabri, Wauchope, Sydney and Young. During those years, I learned which clouds preceded sleet and snow, thunder and lightning, droughts and floods.

"Don't worry, Pet, it's only the clouds banging together," my father would tell me during storms. He himself was especially sensitive to thunder and lightning. They seemed to be precursors to his mood changes and darkest imaginings.

Clouds 'bring drama and variety to our skies', Gavin Pretor-Pinney observes in his *Cloud Spotter's Guide*. "I think they're one of nature's most poetic displays."

I agree. So I'm into cloud spotting these days, too, especially during our long Australian summers, with their big blue skies, powerful thunderstorms, sudden squalls, and cold fronts, brilliant sunrises and sunsets and those lazy changes that turn summer into autumn. The clouds associated with these phenomena puff, stretch and dance their way across the sky to the delight of all cloud spotters. Often, these displays are most dramatic at night, when they are backlit by the Moon and stars. Yes, I wax lyrical, but this is what happens when you become a spotter!

There are two sides to cloud-spotting, I've discovered. The first is rational: what type of cloud is that and what weather does it bring? In inland NSW we are most likely to spot big, dense, white Cirrus, Cirrostratus, and Cumulonimbus clouds. And let's not forget Contrails, those lines of condensation that form behind high altitude aircraft. These new clouds were first observed during World War I but cloud spotters don't take them very seriously. To us, they are unwelcome outliers of the cloud family.

The second side to cloud-spotting is emotional: how clouds make me feel. In this context, contrails are especially affective, because they dissect the sky into little bits. They're a symbol of so much that is wrong with our world, and their impacts on my moods are cut and dried! Yet I can't help wondering where contrails drift to? And what destinations, what adventures, are the people in those aircraft flying to or from? I gaze up and long to jet away to somewhere exotic like that. It's very unsettling.

My diary entries over several months in 2018 reveal a rambling mix of cloud-spottings and a similar mixture of emotional responses to them.

January

> *The morning walk is grey, very dark and cool. The clouds shift by mid-morning to show blue sky and sunlight.*

> *I decided to clean a filing cabinet in the study. Such an accumulation of old photos and documents evoking mixed emotions over past losses, events, people known and loved and some pride in written documents.*

Did I really write that? It's bloody terrific! I certainly had a talent for facts, although they're somewhat boring.

I'm closing doors to the past; they're closing but are not quite shut yet! The clouds have appeared again. White and speckled across the sky. I think these are Cirrocumulus. No rain, but some wind in the trees now.'

The sun comes out and hits the iron cladding of our house. I can hear the crackle of its expansion. Not a lot of birds around but I can hear them in the trees. The clouds have a thin rippled look by mid-afternoon as they slowly cover the whole sky—a stillness, then a soft breeze. The clouds are breaking up now, a marbled effect in the sky to the north.

My morning walk generates energy for the day. Clouds in the east, waning Moon, clear sky after a hot night, wisps of white cloud starting to appear mid-morning. These could be Cumulus. They appear to develop on clear sunny days when the sun heats the ground below. They'll disappear towards evening. Oh, and housework! How I hate it. I must finish as soon as possible. I have shopping and a manicure after lunch. The sky is covered with wispy white cloud. Heat radiates from the asphalt on Main Street. A hot, relentless sun, but curiously, I am feeling pretty upbeat. Lots of energy, chatty and hungry!

Mid-February

The cloud looks like Cirrus, another hot day but with a slight breeze. Have had problems with my email address. My service provider tells me my technology has failed. Bloody oath it has! Seems an error I reported has not been attended to. I am therefore left with a confusing mess. The service department tells me my problem could be fixed by end of day. I live in hope. Another change of cloud appears, and the sun is obscured, which gives us relief from the relentless heat. Everything is so dry. I hope no fires start. So far so good for this month. Found a very dehydrated coriander plant. Bugger!

My morning walk has been glorious. The clouds at sunrise were tinged with brilliant apricot. I think they may be

Cumulonimbus. These clouds herald lightning, heavy rain, flash flooding and gusty winds. What's the old saying? 'Red dawning, shepherds warning. Red night shepherds delight.' The kookaburras are laughing. Will it rain? Kookaburras are always a good sign. I feel hopeful.

And it does. The storm hit in the afternoon, and its strength and ferocity were frightening. And the wind. We just had to sit it out, wondering what the consequences would be. The clean, fresh smell afterwards was wonderful.

Today the wind was strong and cold, and there were few clouds in the sky, but by evening, the wind had dropped, and I could see clouds sweeping across the sky as the sun went down. They could be Cirrus, small patches of cloud mounds, but their colour is peach perfect. It might be a fine day tomorrow and without the wind.

End of February

A very grey morning before the sun came out. Grey days turn me off. The clouds are Stratocumulus and they spread across the sky, low, puffy, grey and white with bits of blue sky in between. I expect a cloudy day but very humid. Did some shopping mid-morning. It felt good after so many hot days.

Have just had a blackout in the house and lost some of this diary entry. Didn't save it soon enough. Will need to start again. If anything makes me rage, it's this. My mood is uncompromising. Also, I am very unhappy about current federal politics. I fear there is more to come and can't believe it will be good. The media is unrelenting. It aims to destroy, not report and they do so because they can.

This afternoon, Cumulus clouds describe how I am feeling. The tops are rounded, puffy and white as the sun hits in late afternoon, while the bottoms are flat and dark. They seem to be a metaphor for what is happening—people promoting truth and justice (the white rounded puff clouds in the sunshine) while having a dark hidden agenda. I am so uptight about it all. Calm down, I say, just calm down!

Travelled to Orange this morning to take delivery of a mulcher. A very grey morning in Orange. Clouds were Cirrostratus. White and grey patches covered the sky in large rounded masses. They can signal thunderstorms later in the day or cold fronts. In this case, I think it'll be the latter. But hot and humid when we arrived back in Grenfell. My wrists were aching. Maybe cold temperatures are coming. They do. I felt stiffness and pain in my shoulders the following morning with temperatures down around ten degrees. I long for this sun to return. I don't do cold weather very well.

Today is the Virgo Full Moon. A bright, sunny day. No clouds whatsoever. I feel really good. Energetic, helpful and needing to go forward with certain of my goals. Excited, because the day feels rich with promise. By mid-afternoon clouds have appeared. Could be Cumulus, round, puffy and white when sunlit. They'll probably disappear by evening. The afternoon has been produc-tive. Found material for my story and feel a sense of satisfaction that at last, I am on the right track.

March

Morning walk was cool with the promise of a fine day. Cirrus clouds were appearing by 8 a.m., white puffs breaking up in all shapes and sizes with some grey undertones in larger masses. I can see more clouds moving very quickly in a race to join these masses.

I feel energetic. This is a new week. What will it bring? The sun shows briefly, and the clouds are massing in the east. Looks stormy. Radio warns of severe thunderstorms, hail, flash flooding in the northern areas of the state, but it remains dry here. We've experienced temperature changes from cold to hot, to moderate, and back again but with sunny skies. Slight breeze and the

faintest of clouds from the south, white cirrus with blue-white
sky. I call these "white-hot days". Days that have to be endured.
No good complaining.

The evenings are the same—plain, blue skies criss-crossed with
contrails. Oh! They irritate me, cutting up the sky.

The weather has not been good for early tilling or sowing.
Ground just too dry. Native vegetation is beginning to suffer, too.
My native trees are wilting because water is limited. Long-range
forecast indicates maybe some rain at the end of the month. For
now, it's a case of be patient, wait and see, ad nauseam.

A wonderful poem by Australian poet Judith Beveridge echoes my frustration about the endless days of summer that are drying up our bush souls. It's simply named "Clouds" and describes the effect the lack of clouds can have on our emotions. She vents her frustration on "endless 'highfalutin' blue skies," and wants the blue skies to "stop their rhetorical grandstanding". She wants them to be "common as whisked egg-whites, or mashed / potatoes on a plate. Clouds that by dusk are the colour / of fish-gutter's gloves. Wage-earner clouds working / like Lanarkshire bog farmers, or Mongolian horseman."[2]

So do I.

Gai Lander lives in Grenfell with husband Stephen. She worked throughout the Central West in the areas of health, social welfare and aged care for most of her professional life. Since then, she has been involved in the Henry Lawson Festival, cultural and local native flora initiatives. More recently, she's been active in art and creative writing. She prefers to work on stories that have "an edge" based on her life experiences.

---

[2] Judith Beveridge, Clouds, *Australian Poetry Journal,* vol. 5, issue 1, July 2015. https://apj.australianpoetry.org/issues/apj-5-1/poem-Clouds-by-Judith-Beveridge/, last accessed 5 September 2019

Phil Leman

Kingfishers

Flashes of azure
Sunlight pierces the darkness
To let the birds through

Phil Leman is a retired schoolteacher who lives in Warren, New South Wales, and doesn't play golf.

Carly Lorente

Celestial Highway

Please let her be dead.

Seven kilometres out of Wattle Springs, in broad daylight, she dives out in front of me. Her heavy body bounces off the bumper and shattered glass fills the windscreen. My dusty wagon moves in slow motion, skidding across the tarmac until it finally comes to a halt. It takes a second for me to realise that the loud, heavy breath that fills the barren, silent landscape, is mine.

Please let her be dead.

Although the rest of me is frozen, hot tears erupt from my eyes. Surprisingly, these were not the first I'd shed on this road trip. Driving along Waterfall Way, the scenic route from the coast, had felt like travelling through Pluto's Underworld, like I was an abducted Persephone. With my damaged headlights, I was barely able to see what was ahead of me. Under this starless night, a deep, terrified wailing had begun somewhere deep inside me, and thoughts of my divorce, the end of my thirteen-year marriage a few months before, filled my head. The pain in my chest rode shotgun.

When the dark skies finally opened up at Tamworth, the bawling stopped. I fell asleep beside the road with Mars shining brightly through the car windows.

Now, getting out to assess the damage and look for the roo, I find that the only thing left of her is a piece of sandy, grey fur stuck to the battered bonnet with a splatter of red. A bloodied trail mark leads back into the bush. I want to say that I get in the car and drive around looking for her, to put her out of her misery—if nature hasn't already done so. But I don't have it in me to help her. Instead, with the front end of the car totalled, and the smashed headlight bulb dangling, I keep moving on shakily towards Condobolin. When I pass a giant eagle a few minutes later, I urge him to head back and take care of the roo.

Normally, it's not my nature to just let things die. Maybe I'm getting better at it. A bit like kangaroos who can't travel backwards because of the shape of their hind legs. It's often difficult to connect with others emotionally and physically, and yet the roo and I had collided effortlessly! Jung might say it was synchronicity. At the next servo, I peel off some of her fur from the car, put it in my purse, and secure the bonnet with duct tape.

I had thrown the swag into the car the day after I received an email announcing that, more than a thousand kilometres away, the Wiradjuri Study Centre was hosting the first Condo SkyFest. It was the day Jupiter moved into its home constellation of Sagittarius, the sign of adventure, mysticism, philosophy, and spiritual quests. I grabbed the bush percolator and oracle cards, dropped the kids off at their Dad's, and burned out of Byron Bay.

The Moon was barely a sliver in its new phase in Scorpio. We astrologers would interpret this as intense integration of shadow, intimacy with self, endings, grief and, as I was to discover, death.

Pluto, the mythic character most associated with Scorpio, has been sitting on my ascendant all year and will continue to do so for a few years yet. The Ascendant, from an astrological point of view, is the value of the ecliptical geocentric longitude (aka celestial longitude) at one's time of birth. When a planet that archetypally represents death, rebirth, and transformation passes through it, your entire cosmic landscape is expected to never be the same.

Before I was a mother, I road-tripped often. Pilgrimage has taken me to the ley lines and Cathar Country of Southern France, to Mexico's Virgin de Guadalupe, and to the dusty backtracks of the Ngaanyatjarra Pitjantjatjara Yankunytjatjara (NPY) Lands and Uluru seeking support from the earth and stars. Now, once again, I take to the road for answers. I need to work out how to marry this new role of single mother with the changing woman that I am. I'm hoping that the skies of the Inland Astro-Trail will tell me. Driving away from my children felt so counter-intuitive, and pulling out from the Golden Guitar at Tamworth with a creaky neck early on Day Two made me wonder why anyone would go on pilgrimage! Yet, millions of people visit sacred sites around the world each year, in the hope of discovering parts of themselves they perceive as missing. There is something in our reciprocal relationship with the land that can connect us to lost parts of our inner landscape, but why do we wait until we are suffering to surrender to it? Why do we let desperation fuel our search for answers?

As I drive deeper into the 700 Kilometre Array, the big skies, so aptly named, open up to me. Out here the heavens appear so much bigger, all that powder blue with the white wispiness of clouds on the dry horizon. I feel myself beginning to merge with the landscape. Oh, how I've missed this space, this sacred place where, free of the daily grind, I can so easily feel at one with the land.

Perhaps because I feel no deep cultural roots of my own, I've always been a story stalker, but after the run-in with the roo, I'm wondering if the land is trying to tell me something more. For some, living mythically can be an organic process of exploring the stories of their own cultures, but what of us white folk born on Aboriginal land? Whether my ancestors arrived here by boat or plane, they left or were removed from homelands they had been bound to by hundreds, perhaps thousands of years of history. Their migration changed their lives and it changed, even more profoundly, the lives of the people they displaced. We're all still recovering from this history. My feelings of un-belonging is part of this. If I were born into another culture, I might feel more supported by elders and mentors in this midlife crisis, this dark night of the soul I seem to be experiencing. In another culture it might be recognised as a spiritual awakening, or an initiation into a new phase of life. My investigations into my own Scottish and Irish ancestors exposed me to Celtic languages, and one word, in particular, sucked the breath from me: hiraeth, which defines a perpetual unexplained yearning for a place you've never known. Is this what I'm feeling? *Hiraeth*? A mytho-poetic sense of belonging I can never experience?

With bumper rattling and missing hubcap, I reach my destination. I arrive at Condo SkyFest on dusk. I'm nervous to announce my profession at a place where scientists might be present, but it's quickly revealed that I needn't be; the cultural astronomy we have gathered for is about decolonising space and embracing ancient stories as a bridge to a sense of belonging.

As Above, So Below is a common phrase in the Byron Shire and I am relieved to learn that Wiradjuri and Gamilaraay people also share this understanding about the reciprocity between Heaven and Earth. Like the early astronomers, I see the constellations above as more than pretty twinkling lights. They offer clues about our place in the cosmic web we are all part of, this merging of matter and mind.

"The universe is made of stories, not atoms," the poet Muriel Rukeyser told us, and I agree. That night, as we huddled together under the telescopes, we were immersing ourselves in the universe of stories. They were there above us, imprinted upon the sky. The celestial emu who rises and squats on his nest according to the seasons. Orion who's chasing the Seven Sisters of the Pleiades star cluster. And the Southern Cross that's a heavenly gum tree where, some say, the spirit of death lives. The two bright stars that point the way to this tree are sulphur-crested cockatoos with eyes of death. They nest in the tree then take off into the sky to remind us that, here on Earth, death is part of life.

Someone in the crowd comments that Condobolin is a special town because you must go out of your way to reach it. There's little reason to pass through, it's true. All of us visitors have made great efforts to be here. It's surprising how far we humans will go for a sense of belonging. We are black, we are white, and there is much to separate us, yet, by all being here, we prove that it doesn't take much to bring us together when we are willing to join the dots, the stars, the spaces in between. I fall asleep at my camp on the Lachlan River feeling the most contented I've felt in a long time. I sit under a gum tree the next morning to let the stories sink in, these building blocks of the cosmos. The sulphur-crested cockatoos and pink galahs are squawking all around me. Warwick Thorton's brilliant film *We Don't Need A Map* comes to mind. In it, he examines the place the Southern Cross constellation has in the Australian psyche. For him, it is becoming more like a Nazi swastika. I, too, have seen how this symbol fuels racism, and an especially toxic nostalgia, perhaps even a touch of hiraeth.

On the second day, I ask Jamie, a Galari Bila Waga Dhaanys leader, if his mob has a mythic character like Persephone or an Underworld. He tells me a story about the dark phase of the Moon. The Sun shines every day, and everyone rejoices, he says. But soon the Moon starts to get sad. Jealous. He feels that no one is paying him any attention—so he wanes. Every day he loses a part of himself until he becomes invisible. The people cry because they miss him. They cry so much that their tears fill a well-known lake. He realises then that they love him, so he makes himself visible again. And slowly he grows into fullness.

I love this story for many reasons. It reminds me to accept the cycles of sadness and light we humans go through, our waxings and wanings. And to not panic in times of sadness, because the light will return. Embracing myth in this way is to acknowledge our human complexity. Not only does it allow us to distance ourselves from human behaviour we find confronting and ugly, such as jealousy or grief, it teaches us that if we are able to accept the actions of imperfect mythic others then perhaps we can also accept and love our own imperfect selves. Telling such stories is a bit like drawing mud maps on the ground to show how hope can lead to transformation, even when life appears murky.

I am so grateful for the belonging and grace that Condo's First Nations people have so generously shared with me at SkyFest.

As I make coffee on the camp stove on my last morning under the Western skies, death again confronts me. A sign that my pilgrimage, for now, is over. Sitting on the warm earth, I recall the following dream from the night before.

My mother forgets to put the handbrake on my car, and it rolls downhill causing a collision with two other parked cars. Mum gets out and blames others for the damage. She argues with one of the drivers until he agrees to pay for the damage. Later on, when she is in the shower, he breaks into the house and kills her.

After a quick pack up, I drive slowly out along the Condobolin road, and my mind wanders to Sufis and their *mundus imaginalis* traditions; how it is not us humans who do the dreaming, but us who are dreamed by the land around us.

I had begun this 2000-kilometre road trip mourning the loss of my old life and fretting about the new way I must now mother. As in the dream, part of my inner mother has had to die, and maybe the part of me that blames others for things that have gone wrong in my life have also gone.

The dreaming track I drive from the Central West of New South Wales back to the Northern Rivers is a big one, in both earthly and cosmic terms. My own story is merging with the landscape. The land has heard my inner tension. I have left my skid marks on the tarmac and have birthed a songline of my own, my tears and dreams mixed with the blood and fur of a dead kangaroo who, by now, has returned to the earth. Such mythmaking connects the land and our individual psyches, and symbolises Earth's power to speak to us. I can see how I needed to physically experience this story of death to grow, how the land has perhaps helped me.

If I am not yet convinced that the veil between dreaming, humans, land and myth is thin, that we truly have emerged from the Earth's imagination acting through us, then a few kilometres down the highway, when I spot Seven Sisters Ridge near Yarrabandai, I most definitely am.

The sight of this ridge, a sacred site that mythically connects the Pleiades star cluster with the land, makes me feel many things. My own ancestors, the Celts, associated the Pleiades with death and mourning. My first introduction to this dreaming was pre-motherhood, when I lived and worked on the NPY lands of Central Australia. "Before there was the word, there was the land, and it was made and watched over by women," mythologist Sharon Blackie wrote. "Women: the creators of life, [are] the bearers of the cup of knowledge and wisdom, personifying the moral and spiritual authority of this fertile green and blue earth."

I sit for a bit at the foot of this ridge on Wiradjuri country with my eyes shut. The voices of the Pitjantjatjara grandmothers singing up their Seven Sisters fill my head. I feel the red dust being kicked up by our feet as we dance bare-chested, painted-up in star lore patterns with ochre, under desert skies on the other end of this songline. I acknowledge that some stories are not mine to tell, yet I will always listen. Although I can follow my own ancestral bloodline back to Europe, and reclaim those songs, and the sense of belonging they sing me, I must also honour those of the land I was born on, and the blood shed by Her original custodians. Perhaps this is what it is to be born white on black land. My relationships with First Australians allows me to bond more deeply with this country. I would argue that in these times of planetary and environmental distress, this is necessary to ensure the survival of all of us.

Of course, my First Nations friends do not exist for the

benefit of lost white folk, many of whom are more likely to be caught navel-gazing than star gazing. But if we do not seek the guidance of Elders to develop and grow as individuals, how are we to experience a wider cultural transformation, and a more harmonious joint narrative? Tapping into the Earth's story as it wants to live through us is necessary for collective human evolution and cultural maturity. Our power lays in the reclamation of these myths, from all our ancestors. Including the original stories that this nation is built on. What if, indeed, it is this mythical resistance that could restore hiraeth, and cultivate belonging for everyone, with the Earth as our partner?

As I turn off the New England Highway, Kev Carmody croons into my headphones:

> No one's lost who finds the Moon or the sweetness of the wattle's bloom.

I take a last look at the inland stars above, and, as humans have done for hundreds of years, I make a wish. I wish that a sense of belonging be restored not just to myself, but to everyone. That the Condo SkyFest becomes symbolic of a cosmic bridge to not only connect black and white people but to connect us all collectively with the mysteries of the sky. A bridge to a richer mythic life of our imaginations built from stardust and storylines. A truly cosmic reconciliation.

Perhaps it was just a small group of people crowded around a telescope gazing at bright lights millions of light-years away. Maybe certain kangaroos just have a death wish, and a gum tree is simply a cockatoo magnet? I'll leave that for you to decide. But, for me, SkyFest left me with a sense of what is possible if we dream together.

Carly Lorente is a freelance journalist and author, astrologer, filmmaker, and all-round cultural disruptor. She wrote her first yarn before she started school and was given her first camera sometime during the awkwardness of puberty. She can't remember the first time she looked up to the sky for answers but is forever musing on her mythopoetic place which, she guesses, is wedged somewhere between land, psyche, eros, and the cosmos. Occasionally the skies respond. Carly has worked with various groups of women around the world using storytelling as a medicinal tool for individual and collective healing. She lives in the hills of Northern NSW with her two children. www.herstorycollective.com

* all misinterpretations of First Nations stories are my own.

Sarah MacKean

Sic Itur Ad Astra

If I ignore the trampled grass
The ugly grooves and cutting hooves
Of snorting cows
Stamping through soft nomad paths
In the brave new world of New South Wales

If I ignore the shiny coils
Of dung on altar floors
The iridescent feasting flies that spot
These sacred groves
The emptied temples, limbless trees
Blue lizards a blur in the shimmer of dusk
Dust churned up to chase away
Red ochre serpents, dotted crocodiles
Scattering in a burst of rainbow wings
Stories drift unsung

If I ignore the glint and the twinkle of
Metal stars stretched between subjugate trees
Bounded, branding wires that scratch
And dispossess the crisp and disappearing wilderness
And the lives and the meaning before us

If I put aside my pen
If I cover up my map, my compass, my level and my rule
My measuring of the miles
If I pretend I am not with these noisy breathing sweating men
I am not shifting in the oozing heat on a canvas sheet
Swatting at gnats at the end of a day
Surveying the brave new world of New South Wales

If I look away from the ruined mosaic
All the broken lines
And drifting songs
And look up
I am no longer bound on the spoilt dark earth
Tethered to a sun-drowned stream
An oily Styx, kissed by stray Moonbeams
The purple hills on the summoning horizon
Melting away in the flicker of fire

I see
Black swatches of velvet smeared with cream
The smudge of a billion stars
And the brilliance of a wide sparkling ribbon
Forever dancing through an infinite night

I catch in a blink
The arching flash of a star flung off and burning to its end
I think as all we shadows think
Beneath the ever pointing stars
How small we are and fast forgotten
Sic itur ad astra
A shiny scrap of hubris
Thus one journeys to the stars
A poet's polish in a vanishing tongue
Gilded with the flourish of Apollo
Doctor of song

His yellow gleam has no meaning in this burning world
Under heavens never seen from Olympus
I am flattened
By a strange milky sky curdled with stars
Sic itur ad astra

The way of the gods, the priests and the heroes,
The hopers, the hunters, the chasers of stars
Our little ships braving uncharted seas
So many golden dreams and destinies

So much hot desire
To conquer the spinning world
The storms of our stories and the silences too
The marching future scuffed from the dust
Fresh footsteps in the ash of distant lands
Such is the path to immortality
Little, lonely, lost in time
They are none of them even the tiniest stepping stone
To the twinkling river of stars

And yet unconquered
The glittering pinpricks beckon
And one fine night above this fallen garden
A little ship will surely brave the dark and
We questing specks will sail the southern skies

Reflections Of A Former Lady's Maid

I filched a pearl worth seven years.
My lady's pearl, it glowed at me
And flung me to this strange lost world,
Full seven swirling seas from home.
In pastures dry I toil, eyes down.
But come the sleeping hours of night
I gaze at star fields splashed with light.

A twilight wash first fades the sun
And melts hot blue to wilting pink.
Then clouds of silk dust suffocate
The long horizon's lemon trail
And smudge the damask sky with ash.
Beneath dusk's sooty finger mark
I down my tools. So falls the dark.

Black cloth be-flecked with glitter white,
I see my lady's dresser top.
Like little Moons her nail tips shine,
Rich silver splinters floating on
The dresser's mirrored jet-black sky.
The wheeling stars are powder grinds,
Fine flakes of lead in bitter wine.

My lady's ghost face glows at me,
An empty tempting lustrous orb.
In every month a perfect pearl,
She tumbles through the spattered black.
A gleaming cream, seductive sheen –
Oh, what care I her pearl's not mine?
The whole damned sky is smeared with shine!

Sarah "Sal" MacKean grew up in the UK and worked there as a barrister for many years before relocating to Parkes with her teenage son in 2015. She now works in special education at Parkes High School while pursuing her many interests in the town. These include painting, reading, bushwalking, dog walking, weekly euchre at the bowling club, and, of course, stargazing. Since arriving here, she has written four draft novels as well as some short stories and poems. She is currently researching the life and times of a famous Australian for her next novel.

Elizabeth Macintosh

The Untold Truth About "Stargazing Live" (or What Really Happened at Siding Spring)

Excited chatter pervaded the air as the school bus followed the road up Mt Woorut, groaning as it manoeuvred around the tight bends, climbed past the last three planets in the Virtual Solar System Drive (the World's Largest) and shuddered to a stop at Siding Spring Observatory. In the adjacent Warrumbungle National Park, bare patches of orange dirt dotted the steep inclines, and burnt trees stood like black matchsticks on hillsides, remnants of the Wambelong fire that engulfed the area in 2013. Rocky spires stretched skyward, forested gorges folded in on themselves below. The late afternoon sunlight cast shadows on the side of the Anglo-Australian Telescope and coloured its dome in a soft apricot glow.

"Now, students, you must be on your best behaviour," announced Mrs X. "You are representing your school and are lucky to be the first class invited here." She sighed before adding under her breath, "Let's hope you're not the last."

As the eight-year-olds tumbled off the bus, Harry Hartley pushed to the front. A woman was waiting for them.

"Hullo, children," she said. "I'm Mrs ABC."

"Hullo, Mrs ABC," they choroused.

"I know all the planets in order from the sun," said Harry. "Mercury, Venus, Earth …"

"That's very impressive, but we must be off," said Mrs ABC.

"Yes, and I'm going to be an astronomer when I grow up. And find aliens. Do you think there are aliens?"

"No! Now follow me up this path to where they're filming *Stargazing Live*. You can watch from the gallery and later I'll take you out to look at the stars."

The group puffed and panted their way up eight flights of stairs to Level 4 where they huddled along the glass panel.

"There's Professor Brian Cox," said Harry Hartley, pointing. "I've seen him on TV. My father says he used to be a rock star. I know all the planets in order …"

"Thank you, Harry," said Mrs X.

"There's Julia Zemiro, look she's waving. My father says she's h …"

"Thank you, Harry!" snapped Mrs X.

Below them, filming began.

"Welcome to another series of *Stargazing Live* brought to you from Siding Spring Observatory near Coonabarabran in NSW. I'm Brian Cox …"

"And I'm Julia Zemiro. Over the next three nights, we will celebrate the southern night sky …"

Harry turned away and glanced at photographs on the gallery wall before facing the film set. Julia Zemiro was looking across at her co-host who added, "Yes, and once again, we have our Citizen Science Challenge. Who knows what they will find?"

"I know all the …" Harry began.

"Ssh," hissed Mrs X through gritted teeth.

The class watched as filming continued with live crosses to other parts of the country.

"This is boring. I know …"

"Ssh."

Mrs ABC beckoned the children to follow as she led them outside.

"I'm going to take you over to 'Space Gandalf', otherwise known as Greg Quicke, and Kumi Taguchi. You can look through the telescopes."

"I'm cold," said Harry.

"Yes, we're all cold," said Mrs X as they arrived at the viewing point.

"Now children, this is Space Gandalf and Kumi." Mrs ABC held out her hand.

"Hullo, Space Gandalf, hullo, Kumi," the class replied in unison.

A man with long grey hair and beard led them over to the largest telescope.

"Why are you here?" asked Harry. "This is a dumb place. The Trig Point is higher."

"We're here because last time it was so windy up there my beard was blowing sideways, that's why!"

"I know all the planets …"

"Ssh. Don't interrupt Space Gandalf!" Mrs X stifled an urge to cause bodily harm.

"I wasn't really interrupting; I was just telling him …"

"Ssh."

"Now, kids, I'll get you to look through the bigger telescope. Up in the northwestern sky, you might see a bright red giant star called Arcturus. It's bigger than our sun and cooler too, but it gives off about 100 times as much light."

Something moved in the bushes.

"What's that noise?" asked Harry.

"Probably a kangaroo. Now, up there," the man continued, "the bright stars of Centaurus …"

Harry moved to the back of the group. Again, the bushes rustled.

"Mrs X, there's something …"

"Ssh. Listen to Space Gandalf!"

Harry peered into the darkness, and slowly, his eyes began to focus.

A strange little man, about a metre tall, looked directly at him.

"Who are you?" Harry asked. "Are you an alien?"

The little man put an overlong finger to his lips before taking off into the bush.

"I just saw an alien!" Harry pushed to the front of the group.

"Don't be silly," said Mrs X.

"It was probably a kangaroo," said Kumi.

"Eek! Aliens!" screamed a classmate.

Space Gandalf sighed. "Look kids, I really don't think there are aliens here. We're looking for them in outer space."

"But I saw one, a little man this high." Harry held out his arm to indicate height. Finding himself the centre of attention, he added, "I think he was green."

"Green? Don't be ridiculous," said Mrs X. "It's too dark to see, anyway."

"I saw a little green man! I did!"

"Um, that's unlikely," said Space Gandalf. "Now, in the night sky up there …"

"I saw a little green man!"

"Stop interrupting," said Mrs X. "And don't be ridiculous, there's no such thing." Mrs X turned back to listen to Space Gandalf.

"I saw a little green man," Harry grumbled to himself. "Just because no one else saw him doesn't mean he wasn't there."

Meanwhile, just outside the small cone of light provided by the TV crew, a little green man sent a message back to his home planet: "No intelligent life here."

Elizabeth Macintosh is a teacher and award-winning writer from Coonabarabran, the Astronomy Capital of Australia. Although she has published articles in newspapers and national magazines, Elizabeth's main interest is in writing for children. Her tales often have a humorous twist and many are inspired by real events. This year, three of Elizabeth's stories will be published in the Creative Kids Tales anthology. She once climbed Mt Vesuvius and was delighted to find a shop at the summit.

Alice Mantel

Farewell from Father O'Reilly

Greetings to you all in the name of the Lord in this my final message as your parish priest.

Three years ago, close to the end of my useful working life, the Bishop assigned me to this parish as a steady hand after the controversial disappearance of our crucified Jesus.

Unlike some others, I did not believe a miraculous event had occurred, because there were rational explanations for the statue's removal. However, since the recent funeral for that chap was held in this church, we seem to have become a pilgrimage destination.

Let me share an intriguing incident that may illuminate these unusual events. About two weeks ago I came across a mobile phone hidden under the corner of the altar. I thought mischievous children had hidden their phone there so I took it to my office.

Later that evening, when I could not sleep, I had a good look at the phone. Somehow (I really do not know how) when I pressed a button I heard an accented voice begin to speak with sighs and pauses. I cannot recall everything, but this is my best attempt to retell the story told by that unknown voice.

I cannot say how long it was before I came to be Awakened. I cannot say how I came to it. The past had gone, rising like mist off a river. At first, the sounds of the Awakening were pitter-pat like rain on my skin, then a drumming, then a wave crashing over my face. The sounds saturated my waking and my sleeping; humming,

134

thrumming and exploding in my mind. I could not distinguish the words, but their meaning etched itself into my bones.

I cannot say how the Consciousness then manifested except as a gnawing pain. Like the noise, the pain began as a feathery touch until I absorbed the anguish and hurt of everything with no beginning or end.

I remember the moment of the Awakening. My entire body was racked by a blinding flash. I wanted to say something, but I could only articulate distantly remembered words. Forgive them, forgive them ... forgive what?

The Awakening was a crystal-clear moment. My consciousness changed from its dream-like state to vivid, sharp Awareness. The noise drew me out of my frozen being into flesh. My own thoughts emerged, and I was a leaf carried on a wave of sounds.

I looked up to see dark wooden arches and beams of light streaming through darkness. I saw faces—young, old, male, female, talking, singing—gazing towards me. Yet really they did not see me at all. Here I was, suspended before them—a form, large as life, my eyes wide open, but there was no compassion for my suffering. When I moved, the pain in my hands and feet became excruciating.

Time passed. I was left in darkness again. My flesh grew heavy and sagged on the rough nails in my hands and feet. I pushed my hand against the nail until it was free, blood lacing down the timber. I leaned on my feet until the nail fell out. I found a wooden support to bear my weight. More time passed until I could no longer stand, so I slid to the ground. Looking up, I saw streaks of blood on the shaft where once I was pinned. Overcome by the effort, I huddled under a table in a ragged ball. Father, my father, where art thou, my father?

Healing sleep came upon me. When I awoke, I slowly walked outside. This world was familiar, but the sounds and smells were different from the world I remembered. Above, I was comforted by the silver moon hanging in the sky as I had seen it in desert places so long ago. I tried to speak. At first, my throat was tight and dry, but then words began to flow in my own tongue. Oh mother, my mother, where art thou, my mother?

I was cold, wearing only a rag around my hips. Returning inside, I found food and water, and long garments to cover myself. Now it was morning. Waiting under the trees, I wondered why I had come to be here.

I noticed two older men come out of the building, dressed in white robes. One spoke to me, gazing at my face and my clothes as if he had never seen a man.

"Have you seen anyone come out carrying anything?"

Surprisingly, I understood his language and shook my head.

Before long there were people milling around, talking excitedly, asking, "Where has it gone?" "How could it have gone? It must be a miracle." I followed the group inside and they pointed to the wooden structure where I had been suspended. My blood streaked the timber and the floor. I wondered how I came to be there at all.

The group became louder. Men in uniform arrived. One turned to me and demanded "Who are you? What is your name? What are you doing here?"

What was my name? I said, "Josiah. I am called Josiah. I am with these people."

The uniformed man then ignored me and spoke to the robed man.

The crowd became more agitated. I turned to face them and started to speak to calm their anxiety and confusion. "My friends, peace be upon you. Be not afraid; there is nothing to fear from this event. Our Father moves in mysterious ways."

I brought quiet to the assembly before the robed man brought the throng inside and began his ceremony. I joined my voice when they cried out "our Father who art in heaven". Oh, father, my father, where art thou? These poor souls need your help.

I saw that the robed man was a priest who gave guidance to those who now stared at the naked wooden crossbeam. Why is it only now they look at that terrible structure with their hearts open and searching?

After the ceremony, I rose to leave, but the priest stopped me and kindly asked me to stay. "Come my son and share our meal," he said. I went to his home and stayed many days with the priests, working in their garden and learning about their life.

At last, I decided that I should leave and find my own way in the world. I spoke to the priests at their evening meal. "My soul aches to care for humanity that comes into the world alone and departs from the world hungry. I will bring light unto them so they may find their way to their loving Father. You, too, must keep the lamp of love burning."

When I finished, the priests clamoured for my attention. I thanked them for their hospitality and walked into the darkness to face the unknown, reassured that my journey was becoming clearer. Oh father, my father, guide me in this eternal night.

So my time passed. There were many days when I was cold, hungry and alone. I walked through cities and towns, on dusty roads and by shady rivers and began to know my own mind. My dark hair hung over my shoulders, my skin became brown and leathery from the sun, and my feet toughened by long roads travelled. I observed the lives of the many souls I met and learned of their dreams and their calamities.

What must I do, I asked myself. The sound of the voices and the cries continued in my head like ocean waves bringing endless sorrow into my being.

Often I sought solace in churches where a figure hung from a crossbeam. The figures differed but none answered my questions. Who am I, if once I was made of wood? How can wood become flesh and blood? Why am I now here as a mortal man? Why do I feel my life is not my own? Of only one thing was I certain—I was given life to show others the path to their true selves.

I met ignorant and self-absorbed people who denied me. But others followed me, such as Pierre and Mariah.

I was present at a music festival, surrounded by many people when I was confronted by a woman pushing a thin, twisted man in a wheelchair. The man spoke to me, "This body may be crippled, but my brain is just fine. I'm here to dig the music. It's all I've got now."

The woman apologised for him and, pushing the chair onwards, she sighed, "I wish there was more to life than surviving."

I addressed the pair. "My beloved friends, there is more to living than your fragile bodies. You are here to give thanks for your existence and look to the reward in your heavenly life. Let me help you to towards your freedom."

I held the arms of the man and gently lifted him upwards. "I say to you, my friend, be strong in your spirit and your body will be your chariot. You will no longer need this chair."

The man wavered but remained standing and took a few steps leaning on my shoulder. "Be strong in your faith, my friend. What is your name?"

"Pierre," he answered. "Pierre Leroc."

"Pierre, you shall be my rock when the tides of men sweep over me."

To the woman, I said, "And you, lady of sadness and tears, you shall find your faith again. You shall be my messenger when darkness falls."

Pierre felt a new energy in his body. His pains subsided, his muscles straightened, and he told me that joy flowed within him like warm wine.

He spoke to Mariah, "Can you believe it? I can stand again, I have no pain." He took some steps, and she reached out to him, but he brushed past her, "It's a miracle, a real miracle."

The onlookers watched, disbelieving and returned to the music.

Later that evening, the watchers told their friends—"See, there was this guy in a wheelchair who couldn't walk, but now he can. It was a miracle. I saw it myself. He got up and walked. There was this skinny long-haired guy in strange clothes who made it happen. I don't know where he went."

Pierre insisted that I stay with them. Together we travelled in Pierre's old van, going here and there, to places that called out to me.

One suffocating summer's day I suggested we find a cooler place to visit, so Pierre drove to a distant lake. On this Sunday afternoon, the townspeople were picnicking and fishing by the lake. Despite many attempts, no one had managed to catch a single fish. Their families sniggered at their futility in the heat.

I brought out a big net. "Come my friends, let us see if we can borrow a boat from this man," I said, approaching a group of men drinking and cursing nearby.

"I won't let a hippie like you take my boat, but I will take you out if you want," said the man. "You gotta be kidding if you think that net will be of any use. No one has caught a thing all day."

We boarded his small powerboat. Almost as soon as we reached the middle of the lake, the sky darkened and rumbling could be heard. The owner began to turn the boat around to shore.

"Wait, my friend, I have not yet cast my net," I called, throwing out the large net and watching it float and then sink under the water. Minutes passed. The water flickered with silver flashes dancing across the surface until the water was a frothing sea of silver shining in the encroaching dark. I tried to pull in the net, but it was too heavy. Pierre and the owner leaned over and pulled the heaving mass onto the boat. Within minutes the boat was awash with writhing fish.

"What the fuck?" yelled the owner, "What is this? How did you do this?"

He was angry and confused. He had never seen so many fish in one haul. How was it possible?

I smiled, "My friend, James, now we have enough fish for everyone. Perhaps we should return before the storm breaks."

The owner stopped, shaken. "How do you know my name? You're some weirdo, aren't you? How did you get so many fish into the net?"

"Master," said Pierre, "there are too many fish. We must go back to land before we sink."

"Do not be afraid, Pierre," I said. "Our Father has provided for us all. Our friend James will have sufficient for himself, his friends and for all the people who came here today. But we do not have much time because the storm is coming."

When we landed, the townspeople came running to see the catch. They called out in amazement, "We have never seen anything like this. This guy must have some kind of magnet to attract the fish."

They scooped up the fish and loaded their cars and looked around, but the rain was crashing down, and we were already well on our way home.

As Pierre drove, he asked me how I made the fish appear. I looked out of the window for an answer, "I am the instrument of my father. I am a fisher of men. I am the lamp to lead you to your true home. I cannot say how I knew the fish would come. I am only the shadow of the one who has come before."

Later in the darkness, I cried out, "Oh, Father, my Father, why hast thou left me here alone?"

Next day the local newspaper carried the headline "Miracle Man Catches Fish Feast". For the next week, newspapers around the country re-posted the story and photograph of the mystery man and his net full of fish.

Some days later I said to Mariah, "My friends, I thank you for your hospitality and kindness to me over these last two years. But my time here is finished, and I must go to my Father's house."

I collected my things and Mariah prepared food and clothes to accompany me.

Pierre protested to Mariah that she did not need to bring so much, but she continued to pack saying, "Our guest is a man of suffering who has carried us through our pain. Now we must help him with his burden."

In the car, Pierre was worried. "Hey, man, why are you going on such a lousy day? It's pouring, and the wind is ferocious."

"It is nearly the blood moon and I must prepare for the Passover meal," I replied.

"What do you mean?" Pierre asked.

I responded, "You must be strong for the hurricane that follows. Do not grow weary or lose heart. We shall break bread together to celebrate the release of those refugees from wrongful imprisonment."

Pierre parked the car while Mariah walked with me to the great cathedral where thousands of marchers gathered carrying banners and chanting. Eventually, everyone was seated. We listened to the speakers on the small stage under the trees while sharing our meal of bread and cheese.

It was my turn to speak. "My friends, you come here standing in the light of truth and armed with the sword of righteousness. We welcome these wrongly imprisoned exiles, forced from their home by cruel dictators and evil bombs. We welcome them to our land, a place of safety …"

There was a low growl in the background, growing louder. Suddenly amidst screams of panic, the terrified crowd scrambled to avoid a convoy of vans. I continued to speak, standing with two others. We faced the glare of the headlights of three large vans that had stopped metres in front of the stage.

Four men got out of the vans, wearing black hoods carrying submachine guns and shouting "Who do you think you are, you mother-fucking Muslim lovers? How about a little sharia law for you then?" "Let's even up the odds."

There was a blast of gunfire, and the two others dropped to the floor. I remained standing.

One man shouted at me. "Who the fuck do you think you are, dressed like that? Some kind of Jesus freak? You need a miracle to save yourself like you did with those fish."

Mariah grabbed the man's arm, sobbing and yelling "Leave him alone, stop it!"

He swore at her, trying to shake her off. Pierre climbed onto the stage.

Oblivious to the commotion, the leader walked up the stage steps. "How are you going to save yourself, Jew boy? Why are you saving those fuckin' terrorists? You are nothin' but a rat-fucking piece of shit letting these dirty arseholes in to blow up our country." He drew close and pulled out a knife from his boot.

"What now, Jew Boy?" The leader thrust the blade upwards into my chest. I rolled backwards. The leader pushed me hard against the tree. "Got what you deserve, Muslim lover. Is it going to be Allahu Akbar now?"

Laughing, the leader pulled out the knife and pinned my hand to the tree, "Jew Boy, is your God going to save you now?"

My lifeblood splashed over Mariah as she crouched near me. My phone slipped into her outstretched hands. Pierre shouted at my attacker, "You murdering dog."

"Father, forgive them for they know not what they do …" A few words was all I remembered.

I leaned into the tree and my mind drifted back to another place and another time. Was this my life's purpose? Had nothing changed in all this time? There is so much pain and so little love in this world. I have tried to lead them to your eternal light …

Oh Father, my Father, Thy will be done.

My strength is failing, my time as parish priest is ended. Decide for yourself if this tale is true. Trust in the miracle of life everlasting.

Yours in faith,

Fr Jim O'Reilly

Alice Mantel's stories reflect the ambiguity of human nature. As a lawyer, she was always interested in the peaks and depths of her clients' lives. Since her retirement, her short stories and non-fiction works explore the subtleties of being human, often reflecting her social justice interests. Alice's new book about assisting women in retirement will be released in late 2019. She lives in Sydney.

Sophie Masson

Big Sky Country

The sky takes up most of the space here. Walking along the road between the houses, I am dizzy with the sensation of looking up: funnelled like a time-traveller into an exhilarating blue vastness. It is a swaggering thing, our sky; not pretty backdrop, not meek scenery, but overbearing, wild, almost scary. Skyscape is the main game; wildlife rituals and dramas more obviously enacted than on the secretive, low-rolling, subtly shaded landscape.

It is as if the denizens of Sky have permission to defy the colour and noise taboos of the modest, almost morbidly modest, Australian bush. Writers of the bush have often been accused of a dun-coloured realism; perhaps few of them have lifted their eyes from the land and into the gaudy wildness above them. Here are birds in improbable, fairytale colours: screeching galahs in preppy pink and grey; rosellas in an icecream cornucopia of raspberry, lime, lemon, and a wild, chemical blue; riots of black cockatoos, with strident yellow or red tail-feathers; fairy-wrens with blue breasts, firetails carrying their sparking brand behind them; blush-cheeked king parrots, their backs a powdered-soft yet luminescent green. The eye is staggered by the range of it, the boldness of it, the proclamation of Nature's passionate excess.

There are other birds, more suitably attired, yet even these are surprising: the harridan-eyed magpies or currawongs, in their sober black and white, meat-eater sharp beaks open to carol some of the most beautiful bird songs to be heard anywhere; pretty, toy-like crested pigeons in dusty blues, pinks and greys, taking off in a clockwork whir and whistle of wings; swallows darting in soft late autumn air, giving you a déjà vu of spring; the solemn flock of ibis, strutting in the paddock like an Egyptian mural; the kookaburra, with its brisk kingfisher's manner, sitting on the telephone wire in summer, with the brown snake it's just killed dangling off the wire next to it, like a discarded, wrinkled tie.

Sometimes, a group of clouds hangs in the sky against other clouds, like a scrim on a stage, and then you might see the pair of wedge-tailed eagles, who make their home in the painted mountain just to the west of us, soaring in between the layers of clouds, like gods appearing in a Greek play. In the bright blue of an autumn sky—so clear in this high tableland that it rinses the eye—a hazelnut-and-cream-coloured kestrel, which has been hanging steady as a melody for several seconds, plunges suddenly, sickeningly, down to the bleached-blond grass. Meanwhile, a willy wagtail alights on the roof of the greenhouse, waggling smugly to itself, dancing through sheer joie de vivre, it seems, while a string of silvery notes glitters from its throat.

Crows, as Norn-like and gloomily eager for slaughter as their brethren the world over, call querulously expectant portents of doom across a sky that's filling with battleship-grey cloud, while single-minded wild ducks descend in a flurry of neat legs and sleek dark heads, aiming for that patch of luminously green clover that seems to appear so pleasantly for them each year. A white-faced heron rises from the dam and imperceptibly transforms from lanky fisherman to graceful skysailor in seconds. And, in summer, noisy mynahs, opportunistic and wary as sneak-thieves, beat a hasty retreat from the raspberries as we approach. They flap up into the lower reaches of the sky, with the lack of urgency born of dimwitted stubbornness and cunning. They know, as we do, that they'll be back later; we can't keep watch over the fruit all day, and even with the nets covering it, they will find a way …

At night, the drama in Sky Country becomes more one with the land, the edges blurring and smudging, though Sky itself, on clear nights, is black silk pinned and needled with thousands and thousands of stars. Bird-life becomes quieter, more modest: the hunters of the night are by their nature and necessities less like liveried warriors swaggering in the open, and more like hidden snipers. Once, though, we saw a tawny frogmouth, its face not all eyes like an owl's, but all mouth, opening and shutting like a fairground toy's, and its body like weathered wood.

One full moon night, when the restless light bleached land and sky alike, there came an errant magpie in otherworldly gleam, singing a song seemingly composed for the Moon, hollowing out the silent blackness of the night, and making us wake from light, uneasy sleep.

Born in Indonesia of French parents, and brought up in France and Australia, **Sophie Masson AM** is the award-winning author of over 60 books for children, young adults and adults. In 2019, she received an AM (Member, General Division) award in the Order of Australia, for significant service to literature as an author, publisher, and through service to literary organisations. A former Chair of the Australian Society of Authors and current Chair of the New England Writers' Centre, Sophie is also a founding partner and publishing director of acclaimed children's books publisher Christmas Press, based in Armidale.

Brydie O'Shea

Corona Australis

To B.J.C. (almost Ernest Hemingway) Wish you were here Dad, to see I kept my promise.'

This One's For You Strider.

> I once was Sagittarius; two heartbeats worked as one.
> Beneath my seat, walked hot shod feet. Above me rode the
> Sun.

Late Evening, Mid April 2007

Destiny led me to Corona.

The Southern Lights, blood-red in the sky, whipped the cosmos raw. It looked freakish; but believe me, I was convinced of ten dimensions that night.

Lady Luck helped too. She threw in her careworn hand. I mean not every plier-possessing girl was due to drive that pockmarked goat track after dark. I felt I was positioned—in exactly the right time and space—to assert the pressure, to cut the fence, to pull the wire, to free him.

Those were the facts. They were the reasons I heard the ping of severed metal and the reason I turned my face from the launching barbs (even as the plasma spun above me). Corona didn't flinch. Too weak with injuries to generate any fight. It seemed only his eyes showed life.

It was his eyes that pulled me in.

His irises reflected the rays and took me to a place among the stars. I had no doubt there was something otherworldly in the air. Magnetism? Electrically charged particles? It gave me a gut feeling. Maybe it was seeing the Southern Aurora for the first time (a phenomenon that far north that was almost unheard of). But one thing was certain: Corona and I went together. Like needle and compass; like north and south.

"You've done yourself some damage," I said to him. My teeth gritted. My hand hurt, 'til the wire gave. "You must have fair flown into this fence. And one might ask 'Why?' Were you jockeyed by the devil?"

Corona nodded, appreciative of the conversation. Not even horses like to die alone.

I phoned the owner Patrick Ahern and waited. The sky wheeled burgundy. The horse and I waited, as Ahern put on clothes and drove the long paddock to meet us.

I filled the interim with deep and meaningfuls. "What's your take on this, Old Boy? Can things once done, become undone?"

Patrick Ahern didn't think so.

"Broken down racehorse," Ahern said (it came out like blasphemy). "I'm gonna dog you!"

To be honest, he had a point. The horse was a mess.

Then I saw his brand through caked-on blood, and something niggled.

Curious, I asked. "What was his racing name?"

I waited for the revelation but felt shocked when it came.

"Corona," he said. "Corona Australis."

My heart stopped mid-gallop.

"Not Corona Australis, who won the Slipper in '95?"

"Yep. Bastard horse's been no use since."

But there was ringing in my ears. I didn't listen anymore.

> I know the heart of Pegasus; to fly and then to fall.
> But we got high, so near the sky, while in the saddle tall.

Next Morning April, 2007

"Geraldine quit being Mother Theresa to animals. What happened to the horse you went to buy?"

"I came home with this instead."

Jack was not happy.

"Why would you do that?"

"It was like this, I justified. I turned into Ahern's driveway, and the horse was in my headlights. It was fully tangled in barbed wire. It must have careered into the fence full throttle. I bet it'd been fighting for hours. Then I rang Ahern, and he rang the vet. They both came. The vet said the horse could live, but Ahern was going to shoot it. So you see; I had no choice. I had to buy him."

"No choice?" Jack looked aghast. "How much did he cost?"

Hesitating, I said "Two thousand dollars."

All colour drained from my husband's face.

"Well, forget about buying a polocrosse horse," Jack said, his voice rising. "King-Of-The-One-Horse-Sports? Bloody, Poor-Man's-Polo, more like it!"

"I'll play on this horse," I said.

Jack looked incredulous. "No, you won't, or not for a long time. Look at him." He threw both arms up. "Season'll be over before he's well enough to ride."

"Jack, please," I pleaded. "Don't be angry."

Jack shook his head. "How can I not be, Little-Miss-Bring-Home-The-Strays? We can't afford a thoroughbred, especially one that belongs in a dog food tin." Jack raked his hand through his hair.

"This horse was a star once, Jack. Remember Corona Australis?"

That stopped Jack in his tracks. "Golden Slipper winner?"

"Absolutely," I said.

"Corona Australis! That's unbelievable!"

"Almost, but it feels right, especially since his name means South, Jack. Crown of the South."

> Half man, half horse is Centaurus – alike, but poles apart
> When Hercules shot Chiron's knees, its poison tracked his heart.

June, 2007

It took months of nurture and training, but Corona Australis rose from near death.

Still, there were problems: he was too much the flying thoroughbred. He shot straight like an arrow, but had no polocrosse stop, or turn. That he was no sport horse was evident.

I decided to play him anyway.

His first carnival dawned to the wafting smell of bacon, mixed with dung. There were sounds of whistles and clashing racquets and the gallop of hooves on harrowed ground.

I rode on the field to cheers, and almost missed the voice yelling from the sideline.

"Mrs North?"

It was Patrick Ahern. Just my luck—he was watching my game.

"Geraldine will do," I said, none too nicely.

We exchanged shrewd smiles. "How interesting," said Ahern. "Geraldine North, riding Crown of the South?" He looked smug as if he'd said something clever. "You must think you're on a winner."

"No," I said. "I've heard crowns are gruelling when they topple."

"Exactly what happened to your horse. His crown took a dive early on in the piece. Flying me to great heights he was, your bloody horse. Then he ups and gets hit by a truck. Bye bye, racing days."

"Corona still flies," I said, with fight in my voice.

"I'm sure he does, but not enough to win races," Ahern said with scorn. "Only a fool would think they could change his galloper ways. Why you'd consider him for polocrosse is beyond me. Except, I see, you're the type who attracts trouble. His smile turned malevolent. "I'm even gonna prophesy that you'll be on your knees the next chukka." He laughed. "I'm prayin' for you girl."

It was true. Ahern *was* praying. The rumour mill was chaotic with that little gem: the fact that Ahern was running from time, and turning to straw-clutched Gods. It took some believing, except for those of us who knew him: just another alcho with liver disease, too scared to die.

I didn't fall, but the game was appalling. Corona and I played No 2. The horse had ballistic acceleration, but I could not turn or stop him. We were a one-person band in a team sport that required us centre field. I felt we both lived up to our names that day, North and South: nowhere near the middle, both of us extreme.

> My soul is mixed with Unicorn; one antlered equine head.
> It rained on down, upon the ground. "The Ark has left!"
> they said.

I loved my horse. It was an 'opposites attract' thing. I thought we'd be together all his days. That was before Jack's dad Selwyn's stroke. He needed care. There was no choice, so we moved to New Zealand to help.

That's how I met Joshua. He wanted to play polocrosse and needed a horse. The Word led him to me—word-of-mouth, that is, which kind of figured. Joshua visited people's houses with stories from the Good Book, unashamedly spreading his religion. Word had it that Ahern was Joshua's greatest fan. Which was lucky for Corona, I guess.

After months of circling, catching, bouncing, and half-halting Ahern became impressed with my work. He even did some word-spreading of his own.

"Corona Australis," he declared, "is now a polocrosse horse."

Joshua was a saviour. I could see why people might speak to him for enlightenment. There was a dash of charm about him, and I found myself telling him things I normally reserved for Jack.

"It was celestial when I met Corona," I said. "The Southern Lights appeared."

Joshua peered at me.

"And what was most bizarre," I continued, "was my surname—North. Then I found out Australis means South. It got me thinking that the two of us were like a compass. Especially when one considers the ring used to keep the compass balanced is called The Rider."

There was an uncomfortable silence, while I had time to read Joshua's face. "Are you for real?" he asked, without moving his lips.

I saw the sides of his mouth curl upward. "I'll wager Corona isn't a horse at all," he said. It was an obvious mock. "Corona Australis is more mythical beast, a unicorn, perhaps?"

My cheeks flushed. I'd exposed my heart for Joshua to shoot through.

More awkward silence followed, broken by Joshua, "Don't worry. Corona will fare OK. At least he'll be a talking point. He's definitely got prestige. Just think of me as Noah rescuing the animals on the Ark."

Standing on the side of the road, I watched Corona go. His tail flapped over top of the float door. Then he disappeared.

Joshua only played one game of polocrosse. Corona bolted on the first chukka, and our team didn't see either of them again. The calls, emails and letters I wrote from New Zealand to the preacher remained unanswered.

> Corona Australis you burn; far closer than Orion
> The magnets slipped and compass dipped, and I left you
> behind.

March, 2010

Our first days back in Australia we paid Joshua a visit. But his house stood empty.

"Don't panic," said Jack.

Except I couldn't help it.

We walked Joshua's treeless paddock on dusk, the heat rising out of the ground. A lone, lack-lustre horse, head slumping, stood melancholy and sick. I could see its star.

"Don't be Corona," I pleaded inside my head.

But I knew it was.

I called. The horse looked, and we recognised each other, we who were connected. He whinnied, and started walking, wobbling at the knees.

Jack didn't speak.

I held my breath.

Corona came and rubbed me with his mouth, his breath rank with abscesses and rotten teeth.

I smelt it, and realised things were bad. Real bad. I held his bony head and saw his ribs protruding from his emaciated body.

"I found a note," Jack said. It shone under his torch.

"To this horse's owner, contact the RSPCA office urgently, or action will be taken." There was a phone number. It was dated six weeks ago.

I erupted in tears. "I failed you," I said to Corona.

"Not true," Jack said, holding my arm.

"Yes, true! And what of Joshua claiming religion? God damned hypocrite."

I looked skyward at the diamond crux above. "I swear upon the Southern Cross to get you out of here, Corona. I'm stealing you back."

Then I got the pliers from the glove box of the car. The same pliers I used the first time I cut Corona free.

> Upon your death—Monoceros, I mourned deep for my loss.
> I cried The Sea of Galilea, beneath the Southern Cross.

That evening, March, 2010

My horse had a homecoming.

I managed to return him to his old enclosure; to his old stall and turf. And long after the vet had gone, I stayed and stroked him.

"Come, Geraldine," Jack said. "It's time we went inside."

He knelt beside me, touched my hand. He was warm on my skin, after the horse. It doesn't take long for the dead to go cold.

Jack said something about burying Corona in the morning, but I only half-listened, looking around at the syringe wrappings and empty packets used by the vet.

"I just want to get pissed!" Jack said, eventually.

"That's possible," I said. "I just want the impossible." I looked from Corona to the heavens. The night sky was black.

"You did everything you could," offered Jack.

"You think so? Leaving him with that Almighty-Equine-Murderer? I hope that bastard rots, worse than the teeth he left in Corona's head."

Jack stood up, kicked the dirt. He was wretched too.

The night was torn by crying. My own.

"Joshua was a bad choice," I said. "I knew it, Jack. I should have intervened when he compared Corona to a unicorn and himself to Noah. I knew he was no good for Corona, even *then*."

I was fessing up. Though not sure who to. Definitely not to my husband. "I should have trusted my instinct and taken Corona back before Joshua left with him."

Jack waited for an explanation. He always waited. Accepted I was cryptic.

"In the Ark legend," I said, "Noah took two of every animal."

I turned toward Jack, with teared-up eyes. "He took a pair of everything, all excepting the unicorns. Those, he left behind."

Jack then asked the dutiful question, the one required of husbands at such times. "Why?"

"Because they guarded The Tree of Life. But when the flood came, they all drowned. Hence, there are no unicorns on earth."

It was a sad child's mythology, made somehow real by the horse carcass prone on the ground.

"Legend says that the unicorns became stars," I said. "And that they turned into the constellation Monoceros."

"Corona Australis was always your star," said Jack. (My husband was one, too, in his own right.) "I bet he's galloping halfway across the Milky Way by now."

His words put a smile on my face.

"In that case, Corona will be flying," I said, my face pointed skywards. "He'll be flying, Jack, flying to extremes, all the way with Pegasus."

Finito.
Black Finito

Brydie O'Shea is a beef farmer, in North West New South Wales, and an animal tragic. She is also a wildlife carer and a gardener, activities which are often incompatible! Once, in a far-off time, Brydie was a mounted policewoman, and a New Zealand zoo guide. She lives with her husband Kelvin, horses B.J. Jimmy and Dolly, a cat called Lynx, Jack London the rescue dog, a bevy of bovine beauties, and Big MacAngus, the bull. Brydie has always dreamed of being a published writer, but until now has been bogged down by life.

Marion Packham

Riverbank Dreaming

Lately, I've been remembering my childhood and youth. These memories are part of who I am, but they bring with them bucketfuls of emotions that linger and need to be shared.

Whenever my family and I share our stories, we're transported back to a feeling we call "Home". It's very strong. "Home" is the Murie, the informal settlement a few kilometres out of Condobolin where so many of my older relatives lived. "Home" is Goobang Street, one of the first streets in town we were allowed to live in. "Home" is sitting on the riverbank or floating in the cool water or laying on the trampoline on a clear night with my brothers and sisters gazing at the glorious stars and the Moon, or, more recently, spotting shooting stars and exploring the Milky Way with my children. "Home" is sharing old memories and making new ones with those we love and care about—to keep our stories alive.

I was born in 1969 so have decades worth of my own memories to share. I spent my first twenty-one years in the same house, which is almost unheard of in my community these days. We lived in Goobang Street—where times could be tough, but love was big. Mum and Dad raised eleven children in a three-bedroom home that dad built himself. The house still stands today. My dear mother, a housewife who cared not only for her own family but for the neighbours' children and anyone else who needed help, had the biggest heart and was the strongest person I've ever known. She had to be. Many times, she was unsure whether she had enough money to feed us all. The family got by with what we had and the good old barter system, which is still alive in country communities. A bag of apples swapped for a basket of eggs, or a meal in exchange for a job well done. We didn't have much, but we shared what we had with our large extended family, friends, and others in our community.

That was just the way it was done back then, the way it has always been done in our communities.

We were taught to work hard, play hard, be respectful, listen to our Elders and, most importantly, to see the funny side of any situation, even in circumstances when we felt like we were running for our lives and our heart was beating out of our chests because we were so scared. We always had to be home before dark, and you certainly didn't whistle after the sun went down. Whistling was like signalling the spirits to find you. We still practise these beliefs today and find ourselves saying exactly the same things to our children that our parents said to us. Sometimes we stayed out too late and had to walk down the back lane under a pitch-black sky, the trees closing in on us, no moon and hardly a single star, and that old abandoned house … it was so scary! I was the youngest child, and my brothers and sisters often teased or tricked me. When we entered the laneway, they'd yell "Run", then disappear and leave me alone in the dark. I'd freeze with fright. I'd stand there crying and singing "Jesus Loves Me" to stop myself being so afraid. Eventually, they'd come back for me, and laugh and ask how singing a song would save me!

Goobang Street was good to us, though. Our houses were very tidy and well-kept, with fruit trees and veggie patches in the yards. We took great pride in our homes, both inside and out. But it was more than pride; there was fear too. Our Elders and parents always had in the back of their minds that someone would report us to the authorities. We feared that, even if our houses were spotless, but the washing up wasn't done quickly enough after a meal, say, we'd be called "dirty black", the stereotype our people had endured for generations. It explains why so many of us are still apologetic if anything is out of place when visitors come. "Sorry, my house is messy!" we all say.

Mum always had a big pot of curry or soup on the stove to feed us at Goobang Street. Often the men brought home a sheep for meat, which was very exciting for us kids. They'd kill the sheep then hang it on the old T-bar clothesline in the back yard for butchering. Our brothers used to chase us girls around the yard with the offal and other horrible left-overs, which always meant lots of squealing and yelling. Milk came in glass bottles then, and we'd fight over who got the cream on top. One of our Aunties made the best johnnycakes for tea, and another cooked the best rice puddings and bread-and-butter puddings.

We also ate a lot of wild food—but we wouldn't dare tell the kids at school about that. We couldn't tell them that Uncle had cooked an echidna and it was sitting in our fridge, a whole echidna with no quills, ready to eat; or that Aunty up the road cooked goannas in a pit in the backyard. We'd also eat a lot of rabbit and fish. I have very fond memories of going bush to set the rabbit traps and checking the night lines along the river. Some nights the moon cast such a glow over every tree, leaf and ripple on the water that we didn't need a torch. Our rabbiting and fishing walks were truly magical times, and always accompanied by stories about the Old Ones, and lessons from nature about animals and the seasons: to be wary of the Willie Wagtail as it danced about, for example, because that bird is the harbinger of bad tidings. It warns you when someone is about to pass.

Wiradjuri words were scattered throughout our conversations as something to be treasured and kept amongst ourselves. Cultural practices of any kind could not be shared widely because we didn't want to be seen as being different from everyone else in town.

In summertime, as the temperature began to rise (along with our tempers), we'd escape to the riverbank or the creek. The heat zapped our energy and made us melt like the tar on the road. We would run barefoot as fast as we could to Goobang Creek, or Coogee, as we called it then. Swimming was the highlight of every summer and we would always feel rejuvenated after a swim. The water would bring us back to life.

> Arms outstretched giving in to all I feel, giving in to all the beauty I see. No thoughts, no cares, just the sound of my breathing and the water slipping through my fingers. This is where I'm meant to be.

We'd swim, make mudslides, race one another to the big log, make death-defying swings from a rope into the water, or leap from the trees when the river was up. I cringe now when I think of my children doing some of the things I was allowed to do. Mum would try to get us home at dusk, but often it was still too hot. Sometimes we didn't leave the river until after dark. When we finished eating and having a shower on those nights, my brothers and sisters and I would grab our sheets and pillows and make our beds on the trampoline in the back yard. When I think of those times now— the closeness of my brothers and sisters as we breathed in the warm earth surrounding us—a tide of nostalgia rushes over me.

Mum, Dad, my brothers, sisters, aunties, uncles and my dear grandmothers often shared stories about their life on the Murie, too. They lived near the lagoon there for roughly eleven years from the late 1940s to the time they moved into town. The Murie years were a time of transition from the era of segregation to when we were 'assimilated' and became fringe dwellers. Life was much tougher then. People lived in tin houses with newspapers or magazines lining the walls, and dirt floors that they wet and compacted so the dust wouldn't swirl around when they swept.

Our people lived and worked hard, but, in the stories they tell, everyone greets one another with a smile and shares a laugh. We remember the Murie as a time when our people watched out and cared for one other. Children made their own toys and played knuckles or jacks, or games of rounders. They were resourceful and very hardy. And everyone loved to celebrate special times, like Bonfire Night. Ladies would dress in their finest clothes for an evening full of dancing and singing. The fire burned so big it was as though the flames wanted to leap into the night sky and be one with the stars.

But there were difficult times, too, when families had to deal with trauma, loss and grief. Many just kept on moving to escape the pain. And there were so many boundaries, in every aspect of their lives, that they couldn't cross. Aboriginal people were segregated at the picture theatre, and they couldn't swim at the local pool, for example. This was a time when men could serve their country in the armed forces but could not be recognised as citizens or go to the local pub to share a beer after a long day's work. We need to remember such stories.

But my relatives also tell some very funny stories about those days. About my father and his brothers driving a horse and sulky at a frantic pace along the Murie's dusty tracks, for example, and the women gathering up their children and screaming "Look out, here come the Wighton men!" A story that reminds me of a good old Western movie! Other stories they tell are about my two aunties when they were young. How, one day, they helped Granny with the washing and used too much starch on the sheets, so the sheets were as stiff as boards when they dried! Another time the aunts thought they'd help Granny by making the damper for tea. They did, but it was so hard no one could eat it! They threw it into the lagoon and watched it sink!

These two aunties of mine were the first Aboriginal children to attend the primary school in town. Their older brothers worked hard to help Nan Wighton raise the younger children and went without shoes so the girls could have shoes for school. Pop passed away early, which meant that Dad and his older brother had to leave school at fourteen. Dad couldn't read or write then, but over the years he taught himself. I remember him as being able to read anything cover-to-cover and debate you on any topic. I still smile when I recall him discussing the bible with a religious group who knocked on our door. They couldn't get away from him quick enough! "I'm a Jack of all trades and a master of none," he would tell us. But, even though he wasn't formally educated, I considered him an educated man: a builder, a plumber and contributor to his community in so many ways. As wild as he was, he could watch the ballet on television one day and a game of football the next.

Education was the one thing in our life we could not negotiate. There were days when we didn't even have enough food for lunch, but we had to attend school anyway. We didn't miss a day. Of course, we were not going to take johnnycakes to school even if Mum had made them that morning for breakfast. We didn't want to be different from the other kids who had sandwiches made out of bread from the shops. Sometimes, on those tar-melting days in summer, we would walk home for lunch in forty-degree heat, and Mum would have a curry on the stove! Times like that we would roll our eyes, grab a drink, an apple, and walk back to school!

In the evenings, it was homework at the kitchen table. If I didn't have set homework, I'd have to rewrite my notes for the day. My parents enforced this. And they wouldn't let us leave school unless we had a trade apprenticeship, a government job, or were going to pursue further education. The military, engineering, nursing, education, juvenile justice, the railway: you name it, someone in our big family has worked in that field.

I didn't know what I wanted to do after completing my Higher School Certificate, though. My family wanted me to do further formal education, but, at the time, I wanted to earn some money. My first job was as a shop assistant at a corner store; then I became a cleaner, a cook, a waitress, a secretary and a bookkeeper. My volunteer work helping children with their reading gave me an interest in education, and I became an Aboriginal Education Assistant. And, eventually, a teacher.

As an educator, I rejoice in, at last, being able to openly discuss our history, including what 'Assimilation' and the Stolen Generations meant to us: loss of culture, loss of language, loss of land, loss of identity. We still carry the burden of this history, as experienced by our parents, grandparents, elders and ancestors. And so do our children. Looking back to the past and sharing our family stories link us together and helps us understand why we do what we do. It also puts fire in our bellies and drives us to keep learning and educating others.

It's vital that we hold onto these stories because they help us to be stronger, to ground ourselves, and to connect with our ancestors, our culture, our history, our people, and with Mother Earth. Something that is needed now more than ever.

The stories of our youth, the stories of our elders and our ancestors touch something deep inside us to stir a sense of belonging and shape our identities. These stories connect us from one generation to the next, and by highlighting the struggles and hardships our ancestors have faced, compel us to do well in this life for them and our children. Yes, we need our formal education, but we also need that connection to our past. Our history shapes who we are and gives us steppingstones to direct us on our own journeys.

I still gaze at the stars as I did in my childhood. I think of them as our ancestors' campfires, and imagine our ancestors waving to us from their campsites along the River in the Sky or Milky Way. I close my eyes and see them singing, dancing and telling stories in the flickering light of their fires. That's where I'll be one day too—reunited with my ancestors, my loved ones. Because, in our Wiradjuri language, there are no goodbyes. What we say is *Guwayu*: I'll see you later.

Marion Packham was born and raised in Condobolin, a small country town at the geographic centre of New South Wales. She trained as a teacher but sees herself more as an educator because lessons can occur anywhere and encompass more that can be taught in schools or from books. She is passionate about Aboriginal dance and culture and works with a dynamic group of people to raise awareness, promote cultural activities and traditional knowledge and help young people learn Respect, or Yindyamarra. Her involvement with her local community is her way of 'giving back' to where she comes from and ensuring the longevity of her small country town.

Pepa Paive

As Above So Below

We were sitting around the fire at the camping ground near the old space tracking station in Orroral Valley, near Canberra. The crackle of the burning wood and the smell of burning eucalyptus leaves reminded us of family *vacaciones* back in southern Chile. Mamá used to throw eucalyptus leaves into the fire there too. My roots are in that smell. We loved camping in the bush and had been wanting to do this trip for a long time. Back in Chile, my brothers and I had joined a scout group. Once, we camped in the middle of a forest during a storm; on another camp we had to survive on a single canteen of water. For us, those times were Wonder Years, like that TV series in the 80s: the three of us sitting in front of an old television set, one with a long antenna at the back and a knob on the right side that you turned to change the channels. Mamá used to make avocado on toast and banana milk for us then. Such wonderful times!

It was nearly midnight when we arrived at the campsite, super-dark and chilly, and, above us, an ocean of stars. I felt a tingling in my belly, which happens to me whenever I look at the night sky. It's as if up there is Home and I'm a daughter of the stars.

Papá parked at the entrance to the camping ground, got out of the car, and walked the main gate. It was locked. He tried to open the catch. Mama wound down the car window.

- Mamá: Lalo, (that's Papá's name) all good?
- Papá: (he turned to face us) Yes babe, it's just this gate is locked!

The headlights illuminated the scene for us. We observed in complete silence like we were watching a TV thriller.

- Mamá: (getting out of the car) Look, there's a guardhouse there.
- Papá: True! Is anybody there?

He called out loudly but nobody answered.

- Mamá: Is somebody there?

Her voice was strained and insistent. Again, silence.

- Mamá: Well, seems that we'll have to come back tomorrow morning!

- Papá: Yes, I agree.

As soon as they turned around to come back to the car, the metal chain on the gate moved. Mamá and Papá stopped and slowly turned around. The chain dropped with a clunk, and the padlock opened. We still don't know how! Even to this day, we can't explain it, but whenever we're sitting around a fire together, we reminisce about that mysterious gate.

We were used to camping so setting up the tent in the dark wasn't a problem. It was fun trying to find things in the torchlight, like a game, and the sky ... Mamá said something to me, but I was too hypnotized by the Milky Way to answer her. The stars hung together in small groups like nebulae, soft white with pinches of gold. One especially attracted my attention because it shone intermittently. Soon the tent was up! The next day, of course, we noticed that some of its parts were missing, but nothing to worry about. It was all part of the adventure. We slept well that night.

The following day, by family vote, we decided to walk to the Booroomba Rocks. I loved to run, to feel the wind on my face and the freedom of the immense landscape, so I soon left everyone else behind. The landscape was so open, so different to me that I felt like I was running across the Moon. I carried a little flute in my pocket and played it now and then. My parents sometimes sang while they walked so that we could follow their song.

We stopped for lunch and Dad laid out his map.

- Me: Papá, what are you looking for?

- Papá: Well, I'd really like to find the path to the Aboriginal rock shelter. The guidebook says there are ancestral paintings there.

- Me: *Oh, como en las películas!*

- Chris (my brother): No! *Como en los libros!*

After another hour of walking, we were all tired. This time, I was the last in the line. In the distance, I saw my brother looking at a rock with his magnifying glass. He was concentrating very intently.

- Me: Hey, Chris!

- Chris: Shh, shhh, shh!

He lifted an arm to stop me but didn't take his eyes from the magnifying glass.

- Chris: This is UNREAL! *Muy bacán*.
- Me: What?
- Chris: They're like hieroglyphs. Brown and red colour. Very symbolic.

To me, they seemed like ancient rock paintings I'd seen in Chile.

Chris was a history nerd. While I carried my flute in my pocket, he carried a pocket guidebook.

- Chris: Look at this! The Ngunnawal people used to live here. And look at that drawing, it means …

He opened his guidebook to the page on Ngunnawal art and showed us the drawings.

- My sister: They're the same! We're in the same place as that photo!

How good is it to have a history researcher in the family! We were so excited that we didn't notice the time. The sun would soon be setting.

- Mamá: Guys, I think it's time to go back!
- Chris: Si, but I don't have any more water, and I'm thirsty. That happens to me when I get excited, I don't know why.

We started walking. We had a long way to go. Soon, Chris complained about being thirsty again. He sat down on a rock and refused to go any further.

- My sister: *Pobre hermanito!* Take my water. There's a bit left.

She sat next to him.

- Chris: I'm so tired I can't move. What I'd love, at this moment, is a super big bottle of cold water.
- My sister: *Jaja*, dream on!

By now, the three of us were sitting on the rock gazing at the sunset. It was very bright and yellow. I hadn't seen a sunset like this before. Every minute that passed made me feel more connected with the universe. This landscape was desolate but so peaceful, and that sunset …

- Chris: *Listo!* We've had a rest! Let's keep going now!

We stood up, and there, at the side of the path, were two bottles of water just like they'd come out of the fridge.

- My sister: Are you seeing what I'm seeing?
- Chris: *Siiii!* (his voice was shaking)
- Me: Maybe Mamá and Papá left them for us.
- Chris: I don't think so. We've been walking for hours without seeing anyone else and these bottles are brand new.

We looked around. Our parents were still walking, but at a good distance from us. And we could see no one else who might have left the bottles.
- My sister: What are you doing?
- Chris: I'm just taking the gift that someone put in our way.

He drank deeply.
- Me: This is very weird. Let's get moving.

My siblings were as unnerved by the water bottles as I was. We soon caught up with our parents. They didn't hesitate to believe in our story.

The night was embracing us, the first stars were appearing, and the Moon was very close. We walked briskly in single file along the path, each with our own torches, our own thoughts.

From time to time, Papá shone his torch around to check that we were all safe. Mamá walked at the end of our file to protect us. Suddenly, we heard steps behind us, like someone running.
- Mamá: Did you hear that?
- Papá: Yes!

He shone his torch around, but no one was there.

We kept walking.

The sky became darker, the stars brighter. The Moon in stillness, the shooting stars like soft fireworks. Such a contrast between the intense dark bluish-almost-black sky and the white tails of the shooting stars. In Chile, we visited the Elqui Valley to experience the Rain of Stars, a celestial phenomenon that occurs several times a year when showers of dust particles disintegrate in Earth's atmosphere. The Rain of Stars was medicine for our souls. Now we were experiencing a similar medicine on the other side of the planet.

Mamá prepared a big pot of tomato soup, with, of course, avocado, when we reached our campsite. We toasted thick slices of bread on the fire, then sat back to enjoy our meal and share stories about our walk and its mysteries. I didn't feel any fear about what had happened, even about the water bottles. My brother thought the souls of the Ngunnawal ancestors were guarding this place and looking after visitors. Or beings from another planet could have visited us, he suggested. I couldn't explain what we'd experienced, but surely there were other beings up there contemplating us like we were contemplating them. As above, so below. Everything is connected, and all that.

I felt such indescribable joy that night. It was one of our best ever family *vacaciones*!

Pepa Paiva is a Latin American artist living in Melbourne. She arrived in Australia three years ago to improve her English writing skills. Pepa's current creative projects include a screenplay about blindness and vulnerability and her first theatre production. She is very passionate about humanity's relationship with the night sky and the beauty of the cosmos. This is her first publication.

Helena Pastor

Mother is watching

Mother is watching
for signs she knows too well
she suspects, of course
pushes thoughts away
not this boy
not her youngest
not the last-born of nine

No, this gift is not one
she wishes to share
this gift is too strong
this gift is to fear

'Do you want that?'
A fork is poised above his plate
so lost in dreams forgets to eat
so busy dreaming he can't sleep
at night he flies
to other worlds
more interesting than here
he thinks nobody knows

But mother is watching

Walls are shifting
ceiling lifting
he is shrinking
room is growing
what is happening?

Why is he vibrating?

Make it stop now
quick, drink water
lots of water
drink it down
but he can't reach the tap
and now he's crying
'Am I going crazy?'
he asks a trusted sister
she looks at him strangely
and he asks her no more

This gift is not one
he wishes to share
this gift is too strong
this gift is to fear

He belongs
to a tribe of nine
at night, he has to choose:
be in the lounge, be with his family
or wait in his room for escape?
He chooses to wait
a noise in his head
a shift in the air
a movement, a sound
and he's away
free

Always reading
hardly sleeping
what is going on?
He dares to ask another sister:
'What is happening?'
'You're a freak,' she tells him
'Just be normal,' warns his mother
'It's not safe to be who you are
You must fit in.'

She is always watching.

This gift is not one
she wishes to share
this gift is too strong
this gift is to fear.

No, this gift is not one
she wishes to share
this gift is too strong
this gift is to fear.

Helena Pastor lives in Armidale and is the author of *Wild Boys: A Parent's Story of Tough Love* (UQP, 2015). Her writing has attracted two Australian Society of Authors' Mentorships, along with residencies at Varuna Writers' House, Bundanon, and Booranga and KSP Writers' Centres. She is also a songwriter and lyricist. "Mother is Watching" is one of sixteen songs she wrote for *Lullaby and Lament – A Song Cycle,* a collaborative project with composer Christopher Purcell.

Simon Pockley

In Place

I'm a mountain person. I like solitude, cool air, high vantage points, rock under my feet, and the feeling of being enclosed by a broken skyline of crags and cliffs. Some people have no sense of where they belong; others fit in anywhere. Flat-country people open their arms to big skies with low horizons; coast people reach for headlands. A few feel so out of place that when they have a surrogate home, they prominently assign a name to it from afar as if this public assertion has the power to call up the place they long to be.

Common ground is that sense of belonging to a landscape that's found its way inside us; places that make us feel alive or complete. Connection to country can prompt questions about Indigenous authority, but it's unlikely that this sense of belonging is inherited through any ancestral affiliation or childhood imprint. I grew up near the sea. The central desert country spoke to my father. My mother spent 70 years refining a garden that resembled a forest clearing. Yet the mountain landscape I look out from has always been inside me and is inseparable from any sense of self I might have. I can't think of anything (apart from love) that, when found, has such a sense of rightness or enchantment.

Life can take me away from my mountain home in the Warrumbungles, but it so closely resembles my mental or internal landscape that I can call it up any time and glide around like a spirit-being without concern for gravity or the continuities of time. In this unbounded space, I'm mostly at rest in some reverie, half-awake to murmurs of memory. Sometimes I'm working through the practicalities of habitation. Even with my feet on the ground, at least one layer of my perception is moistened with the residues of these imaginary visits. But the gains and losses of change mean there are certain places I can only return to in spirit.

If I could take you back in time to a sandstone rise in the middle of the mountain pass that is my home, we would move aside a few sticks and lie on our backs in dry leaves among scribbly gums. This little community grows nowhere else around here, and these trees are people to me. The slender stems that reach into the high canopy are gently swaying. Clusters of curved leaves are nodding and glint against an impossibly blue sky. The wavy under-sound of leaf-slide smooths all separate movements into an over-rhythm of connected motion. Focus on the framed sky-shapes in the canopy, and a slow-as-growth, insistent but acquiescent branch-dance appears. Branches curve away from each other as if mimicking please-go-before-me gestures—never colliding, always reaching for that last fragment of sky and with each change of direction—the symmetry of a matching response.

Further down, among bare stems and trunks, it's clear that some of the younger, slender, single-stemmed trees have leapt through the half-light and grown so vigorously that they've shot straight up to secure their patch of blue. Some (I'm guessing they're female) are overtly sensual with vulva-like skin folds where limbs fork like spreading legs. The lower regions of the smooth grey trunks display, like tattoos, the looping calligraphic tracks of moth lava that give the trees their name. A couple of very old trees with gnarled, calloused knobs are drooling with a bitter-tasting iron-red resin.

The dappled light that reaches the bark, twig and leaf-fall floor of this grove supports just a few sticky daisy bushes, orchids, cycads and occasional clumps of soft, slender grasses. A pair of crimson rosellas arc through. There's a gentle airiness, soft rustlings, and the wafting scent of eucalyptus.

For all their varied ages and columnar stability, the scribbly gums cycle through a dramatic and collective seasonal transformation. It's an event that usually begins with the rising summer heat in mid-to-late December. The cool, smooth, grey bark of the bigger trunks suddenly begins to crack. Bark peels back into ringlets and starts to fragment into bright red shards as if some inner oxidising furnace had been kindled to the point of explosion. With each breeze, flakes of bark-fall settle around the bases of the stems so that the contrasting red litter makes the bright yellow trunks appear like young sprouts. Rain invariably follows, and the yellow colour of the bare trunks washes off and bleaches to parchment, upon which, new moth calligraphies are revealed. Then, almost imperceptibly, over the following months, the trunks mottle grey. My strong affinity with this annual transformation has given it a name—the great shedding.

Far away, in childhood, I made drawings of these yet-to-be-seen mountains. The form of the landscape became such a persistent and recurring expression of longing that in 1965, when I was 16—when I first saw the Warrumbungles in the distance from a friend's father's property, Mundroola, near Coolah—it was with a sense of recognition rather than astonishment that the same pointed mountains I'd drawn were right there in front of me outlined on the horizon. I felt as though I'd just stepped into my future. I turned to my friend and announced that as soon as I could legally drive a car I would go to those peaks on the horizon and live there. But it wasn't until my early 20s that I encountered Harry Harris, a Coonamble cattle dealer, who sold me Wheoh—also known as Bugaldie Gap—a left-over portion of a larger property that had been incorporated into the Warrumbungle National Park just to the east of Siding Spring Observatory.

A couple of blurry photos bring back fading first impressions: alert, wandering around in light, misty rain and climbing the sheer wall of rock that marks the southern boundary—a knife-edged igneous dyke known to some as Scabby Rock and to others as Uncle Ernie's Rock (after the remarkable (legless) Ernest Blackburn, an early owner of Wheoh). I'd like to say that I fell in love with the place immediately, but the remnants of ring-barked trees and scattered evidence of failed agriculture gave the land a neglected feel. It took time to explore, attune, and begin to find my way into its moods and rhythms.

After setting up a camp at the head of the gap, to the north of the scribbly gums, where there was a spring, I set about digging into the ground on the western end of a high ridge from where there was a spectacular view. Equipped with youthful enthusiasm, a pick, shovel and wheelbarrow, I dug right through summer to create a flat space for the foundations of a room loosely imagined as the southern wing of an octagonal tower.

The excavations became all-consuming; as much an uncovering of memory as a descent into the hard ground. In the narrow monotony of such effort, what came at me was probably some artefact of muscle or body memory. It threw up the residual presences of conversations and regrets. I dwelt on my failings. It was a painful, even punishing physical and mental experience. Anxiety came in long waves where tension was followed by release and misery by euphoria. Over time, these waves seemed to stretch out for longer, but the peaks and troughs deepened. I look back on it now as a journey into the unconscious and a self-reliant shedding of youth.

The motion of shedding and discarding pervades the contours of the landscape as well as its naming. The Warrumbungles was formed when volcanoes erupted through an ancient layered sandstone tableland and then weathered away leaving the stumps of domes, spires, dykes and sills protruding from forested ridges and deep gorges.

Only a few Indigenous names remain for these features. My favourite is the unused and all-but-forgotten Tha-a-ma. It's believed to be the phonetic spelling of the Aboriginal (possibly Kawambarai) name for Timor Rock before it was anglicised. It's why we locals still say tImor and not teemor. Tha-a-ma means 'the runaway'. It's worth saying out-loud for its gentleness and, for a moment, it breaks the silence of loss—loss of that vast and indefinite period of Indigenous occupation and voice. It so perfectly and humorously captures the sense of a landscape in motion, of what would otherwise be just a large rock, standing alone, half-way along the road to the Warrumbungles from the little town of Coonabarabran.

Circular stone arrangements, sharpening grooves and some struck tools also endure, but it's hard to accept that the ancestral stories and inventions that once enfolded this place in a greater meaning are irretrievable. I wish that these stories had a life beyond their long-lost affiliations so that we could bring them back as if from a hiding place.

Having at its core a gap, a small pass in a line of bluffs and steep, cliff-sided mountains, Wheoh is energised by passages or pathways for the movement of winds, water, wild animals, shadows, machines, even time. The forces of these movements range from tectonic upheavals, where great sheets of rock suddenly detach from cliffs to extended moments of such stillness that silence has weight. At night, the black broken skyline shrinks away from the glittering expanse of stars as if purposely calibrated to mark their apparent movement.

The rhythm of the day can be sensed as the course of an invisible pendulum. Most mornings the east-facing cliffs of Saddleback ignite and the sunlight floods in to push a line of shadow across the arc of the valley curve—only to retreat again in the evening. I never tire of the way passages fleetingly open as certain parts of trees, bushes and even patches of grass light up, as if illuminated from within. As the day settles and the colours begin draining away, it's the dominant mountain, Bulleamble's turn to light up and glow to crimson before fading into darkness.

The breath of Saddleback pulses out of the gorges to the north, washing over the cliffs as if to the beat of a stormy coast on the other side. But it's the ephemeral puffs and eddies that seem to have a life and purpose of their own. I once saw a tiny white cloud drifting up the edge of the valley when a male black cypress pine discharged its pollen in the direction of a grove of females. During moments of perfect stillness, the leaves of a single tree or bush can suddenly begin to shiver and stream in response to some private ferment. Occasionally, little pockets of warm airbrush bare skin as if radiating from the passing bodies of invisible beings. These spontaneous little unnamed wafts and rustles engender presences that hover on the edge of consciousness. Named and described, they take on substance, character and power as they interact with other raw and unexpected energies persistently leaking through the misty shell of awareness and comprehension.

The clear skies of the Warrumbungles are favoured by astronomers, and the full Moon is usually so bright that walking among Moon-lit shadows is a special pleasure. One night, when I was standing on a ridge below Bulleamble, marvelling at the spectacular line of peaks that step down into the western plains like the backbone of some half-buried creature, I felt myself rising into the air from where I looked down and saw myself located in space and time with a sense of exalted connectedness. It was like arriving at a set of pre-determined coordinates. All time—all movement—converged on that point. It felt like being at the centre of the universe but at the same time dissipated into its fabric.

Many people have reported similar out-of-body experiences. Some invoke supernatural forces. Others say neuroscience has uncovered disturbances or lesions in an area of the brain associated with spatial cognition. Regardless of the source of this experience, it was both life-enhancing and memorable. It was certainly a point from which I began to feel so integrated with my surroundings that any sense of separation belonged to a former life.

A strong waking daydream followed. I wasn't asleep, just consciously dozing …

I was flying (like superman) about 20 metres above the route of the road heading east towards town. I became aware of being on a collision course with someone (male—not known to me) flying in the same way towards me. In slow motion, we both straightened up into a vertical position with arms and legs outstretched. Below us, on the road, a green utility, heading away from town, approached a small open culvert where the road crosses Flaggy Creek. Our bodies collided very gently at the point of our belly buttons. At the moment of impact, a flat stone came out from between us and fell to the ground where it rolled on its edge along the side of the road for a short distance and then flopped into the water of the culvert. My attention was focussed on the stone rather than the aftermath of the mid-air collision.

That was the extent of the waking dream. It was powerful and memorable. I thought little more of it except that I decided that I would check the culvert for the stone the next time I went to town. I was not expecting much. There was nothing particularly remarkable about the stone. It was the same kind of sandstone found beneath the scribbly gums; flat, not quite round, and about the size of an outstretched hand. A week later, when I went to town, it slipped my mind. But on my way home, I was approaching Flaggy Creek when my memory was jogged by a green utility passing on its way into town. By this time, I had driven through the culvert. I pulled over and walked back. To my amazement and delight, there was the stone beside the road in the water just as I had seen it!

My delight was short-lived. As I picked up the stone, I felt a sharp pain on the right side of my chest. The dream-gift suddenly became something I was ill-equipped to manage. When I took the stone home and put it into an old bath used to collect rainwater, the pain immediately went away.

Years later, believing the stone to be somehow malignant, I returned it to Flaggy Creek where, within days, a flood carried it away. Then, more years later, when I realised it was a communication from the land itself, I had no trouble finding it again in the dry creek bed. This time I stored it in the base of a hollow tree, up near the spring in the very centre of the mountain gap. The tree was struck by lightning and somehow twisted around the stone. In 2013 a bush fire swept through Wheoh. The charred stump of the tree still contains the stone. Intact.

The significance of the stone is something I try not to rationalise or explain. I'd struggle to find appropriate terminology that wouldn't diminish the rich layers of associations of a lifetime that now give it meaning. As a physical object, it is in no way remarkable or attractive. But I'm always aware of its presence—the feeling of revolving around it. It anchors me with an elasticity that lets me go but draws me back. I feel fortunate to have such a tangible and powerful introduction to what has become an extended conversation between my unconscious and the energies that have drawn me to this place.

Among the scribbly gums, we shift into the present. It's been six years since the bush fire asserted its ecological rite of passage. Fire was always expected, and will certainly come again. It arrived with a scale and intensity beyond my imaginative capacity to foresee. Heat and flame scoured every surface, shattered stone and left a silent, skeletal world suspended—the bare bones of the landscape.

I mourn the death of the scribbly gums as I mourn the death of old friends and all the creatures for whom this was home. The canopy is gone, and the older dead trees are collapsing, their heavier branches dropping away like the life-promises of epicormic shoots. All around, grey leaning limb-wrack and shatter makes what was familiar—strange. Rubble and ruin litter the space where I'd carefully built a tall house among these trees when I was a young man.

Slowly the grove of scribbly gums is rising again through an almost impenetrable understory of wattle and wild hops. I count the stems re-sprouting from their roots. Each year a new understory species takes advantage of the last and dominates the cycle of succession. It's a privilege to be part of the process of regeneration where past, present, and future are like transparent overlays leaching into each other. I'm now an old man, but I too have begun again. The fire has made me bolder and more assertive—more confident. Birds return. Recently a goanna appeared. Barring another fire, I'm guessing it'll be another lifespan before the canopy closes again.

Bush fires, wild storms, floods, even falling in love, are events that compact to mark moments in the span of a life passing. Some leave traces that can wear into the rounded fragments of memory and story—narratives of endeavour; others leave no trace. I watch one of my daughters braid native grasses by instinct and receive the gift of a rain stone from the other. My son waits deferentially. There's an unspoken language of the wild in us that comes with knowing our place.

My brief human span seems helplessly out of scale with the immense power and volatility of the energies around me, but I am in place here—privately and gratefully enveloped in an enchantment that is continuously refreshed by the restless, bristling adjustments of life on and off the ground—beneath sky.

Simon Pockley enjoys time with his family. He mainly works with his hands but has also worked as a manager, analyst, information designer and academic in the business, cultural, environmental and education sectors. He is a committed environmental activist, living quietly and reclusively in the Warrumbungles in north-western NSW where he is re-building after being burnt out by a bush fire.

Rhonda Poholke

Praise The 700 Kilometre Array

And praise The Breadknife[3] for she is most certainly the Goddess of the Array. She has stood in her stone-hard resolve for 18 million years, has watched the people come and go, the seeds grow into trees, and she has seen the telescopes evolve. Remembers being lava, thrust upon the smooth and boundless blue. The Breadknife saw light spread across the land, and night fall with the ease of a silk scarf around her. The stars were younger then and took their places like obeying children. How they sang and danced and laughed as the first peoples gathered stories for the coming ones; and, for every story, they say the Goddess released an exotic bird to fly the Array. The telescopes will hold witness to all of this. They say there is a spot so secret only the Goddess knows where, that if you happen to stand there and fasten an eye to the starry sky, you can read every single word that has ever been written.

Rhonda Poholke is a poet, photographer, collagist, living in country Victoria. A constant sky gazer, she sees all kinds of things in the shapes of clouds. But it is the fascination of the starry nightscape that flies her into other universes and lets her believe all things are possible.

---

[3] A conspicuous volcanic dyke formation in Warrumbungle National Park near Coonabarabran, NSW. https://en.wikipedia.org/wiki/The_Breadknife

Max Pringle

The night as I remember

I grew up on a small dairy farm at Quorrobolong in the Hunter River Valley region of New South Wales. Mum called me her War Baby because I was born in 1940, the middle child of seven. I spent my first few years with my mother's parents, Nanna and Fardy, who were strict but also generous and loving. Fardy had a rabbit run. He'd sometimes let me go with him to set the traps at dusk and to check them later in the night. Rabbit meat was good food back then. Fardy would also take me out to Tucker's Lane to collect firewood. He'd always know what time it was time by looking at the sun. He'd tell me when to start loading the firewood so we'd make it back home for lunch.

When Nanna and Fardy were too old to look after me we all went to live with Dad and Mum on the farm. Dad had a milk run during the war years. He collected the milk cans from the farmers in the region and delivered them to the Bowthorn Butter Factory at Morpeth near East Maitland. He also butchered meat for the family and friends and kept large areas of land under cultivation with food crops.

We children all helped around the farm. When I was around six-years-old, my job was to get up early, around 3.30 a.m., and bring the cows in for milking. Later on, I'd help to water the vegetable gardens and feed the calves and pigs. We often fed the calves in the Moonlight when the evening milking was late.

On weekends my two older brothers, Darrell and Bob, and a couple of their friends from town, would take me camping with them in the scrub. We'd make a bush hut by tying a few saplings together and cladding them with branches. We left the top open so we could gaze at the stars at night. It was a wondrous feeling back then to hear the older boys talk about the different constellations. The sky was so crystal clear it seemed we could see forever.

Darrell and Bob found different interests as they grew older and that was the end of our camping trips, although not the end of my stargazing. I had a mate called Ray Lomas I met at high school. Ray was from town and would come out to the farm for sleepovers. One time we made two hammocks out of a couple of hessian feed bags and strung them between the trees so we could sleep under the stars. We'd stare at the stars till all hours.

We had no electricity when I was growing up, even though the farm we lived on was only seven miles out of Cessnock. Mum cooked on an old fuel stove and did the washing by hand in an old round tub under an acacia tree. We had to carry all the water we used for drinking, cooking and washing from a tank outside the kitchen. We also had a tank with a tap to supply water to the bathroom. We drew all the water for outside use from an underground well near the back door. Our dunny (toilet) was down the back yard. It was of the can variety and, with so many of us in the family, frequently needed emptying. As for a phone, there were only three or four in the whole district and one of those was at my paternal grandmother's place. She was the local postmistress and lived around four miles from us.

When I left school, I worked on the farm and in Cessnock for a while and then spent three years in Sydney. I started at Greater Union Theatre in Market Street as an office assistant. I wasn't suited for that kind of work, though, so I left and found a job as a process worker. That lasted some three months before I found better paid work in a wool store. I remained there until my brothers came to Sydney to work for the construction company Theiss Bros on the fuel pipe line from the Clyde Refinery to Port Botany. I joined them there. When we'd completed the pipeline I returned to the Hunter region and started up a convenience store. Soon after, I married Dot, a farmer's daughter.

Dot was very family oriented, so when her sister Alma married an Englishman and moved to London, she decided we should visit them. We were away for two years. My sister Kay and her husband Merv ran the business as their own while we were in London.

The hardest thing about London was that we could rarely see the stars for mist, fog, rain or snow. It wasn't a bit like Australia where we could get a clear view of the night sky almost any night.

I worked at White Arrow Services in Peckham. One day one of my friends from work, Harry Hubbling, invited me to see his new fishpond. He was very proud of it. His mother offered me a cup of tea—and remarked that she had lived through the most exciting time in history. She'd seen the first motor cars, she told me, the first aeroplanes, the first submarines, and now she'd seen men walking on the Moon!

I was so worn out from work that I fell asleep in front of the television before Neil Armstrong and Buzz Aldrin stepped onto the Moon's surface—I often did three different jobs, sometimes two on the same day, one during daylight and then my regular job at night—but Dot witnessed it all. A few days later we were talking about the Moon landing and I remembered that my ancestors had planted their crops by the phases of the Moon. Back home, Dot's family was still keeping these moon planting traditions alive.

I'd never been interested in the practice of moon planting, or how our ancestors grew their food, but somehow talking about the Moon landing inspired us to visit Stonehenge in Wiltshire. The site was open to the public in those days so we were able to wander around freely. We were amazed at its size and marvelled at how people were able to build such a large structure five thousand years ago with such primitive tools. We'd heard that the stones were aligned to the solstices in mid-winter and mid-summer but no one seemed to know why. I've heard many theories, including that it was constructed by Aliens, although I doubt that. Another theory is that it was some kind of astronomical observatory. Human remains have been found there too, so it must have been used as a burial ground as well.

I've since learned that Stonehenge was also considered to be a place of healing. People believed that the smaller bluestones from Wales had special healing powers. So strong was this belief that visitors broke off pieces of rock to take home. Australian scholar Lynne Kelly believes the monument was part of a 'memory code', a mnemonic device to help people remember things that were important to their culture, and to their survival from one generation to the next. But the more I've read about Stonehenge the more confused I've become about why people built it.

After our two years in London I returned to my business and became involved with the Scout movement and, once again, the night sky became important to me. I was often called on to explain the constellations and other celestial objects to the Cub Scouts when we were on overnight stays in the bush. That phase of my life ended many years back and, unfortunately, I find I have forgotten much of what I knew then.

Dot always had poor health. Even as a child she was taken to Sydney frequently for tests and examination for an enlarged spleen. And then, after our twins were born, she developed cancer and had to go to Sydney for radiation treatment. I managed to get someone to look after the shop while I drove her to Sydney to stay with her sister, Alma. The treatment was traumatic, but Dot managed to cope with it successfully. This was not the first time we'd been separated because of her health though. When she was expecting our twins, I had to take her to Sydney and settle her into a flat near the King George Hospital at Newtown. After the twins were born, a daughter and a son, she was rushed back to surgery for a major operation to remove the spleen and was not expected to survive. She not only survived though, she made a remarkable recovery. Twice again the cancer returned and she beat it off both times. The cancer treatment caused her heart to enlarge and, at one time, she was on the hospital's list for a heart transplant. Instead, the doctors gave her a new aorta valve.

In 1991 we purchased a property near Narrabri and again became very aware of the night sky and the brightness of the stars in the bush. Our new farm had no accommodation so, in those early days, we'd camp in our van with a piece of tarpaulin stretched between the van roof and the cattle yard fence. Sitting out under the stars enjoying a pleasant meal and cup of coffee was very relaxing after a hard day's work on the farm. It seemed you could see forever in the heavens.

We farmed the Narrabri property together for some twelve years until we discovered that Dot had developed a huge tumour in her stomach and needed another operation. This surgery was not successful and, some months later, Dot passed away. I remained on the farm for several years before selling up and moving in to town. Every morning I would rise around 5.30 a.m. and set off for my daily walk. Narrabri had a relatively clear sky and I would get a good view of the stars overhead before dawn, even with the streetlights. My walks were usually timed to conclude as the sun rose. Over time I got to recognise the seasons by where the sun rose over a certain tree in my yard. I suppose this was my own version of Stonehenge, or my 'memory code'.

Australia's First Nations people also use memory codes and have their own theories about the night sky. Traditional cultural practitioner Steven Booby, a Gamilaraay[4] man from Narrabri, told me that the Milky Way, or Warambul, was the great river in the sky, and that Boolimah, a place behind the Milky Way, was where his ancestors' spirits lived.[5] Steven also showed me Gawarrgay, the Emu in the Sky, at the centre of the Milky Way.

Since giving up the farm I've travelled a lot and gazed at the night sky from many different places. Australia has the clearest sky for viewing the stars of any country I've visited. I've often been called upon to escort tourists around Narrabri and they have been amazed by what they can see on a clear dark night. I've also had the pleasure of taking coachloads of tourists to our famous radio telescope, the Australia Telescope Compact Array (ATCA) at the Paul Wild Observatory, part of CSIRO's Australia Telescope National Facility. The static exhibits at ATCA's interpretation centre explain some of the radio astronomy that astrophysicists are doing there. I especially like the photos of supernovas.

I am well into my seventies now, so I can look back over the many decades of technological advancement I have been privileged to experience. The cumbersome old television that Dot watched the Moon landing on has been replaced by a slimline TV with Wi-Fi and computer functions. I now have a laptop computer with a mobile modem that allows me to access the internet from just about anywhere. Whenever I have a problem with my computer my son-in-law near Melbourne can fix it remotely. Tractors can now be driven by computers, a driverless car is just around the corner, and I have a SatNav system to direct me to wherever I want to go. Solar energy means that we can have electricity almost

---

[4] Also transliterated as Gamilaroi, Gomeroi, Goomeroi and Kamilaroi.
[5] See http://www.aboriginalastronomy.com.au/content/community/kami-laroi/

anywhere on Earth, and we're exploring Space and seeing the universe as never before.

I therefore think it would be safe for me to respectfully contradict old Mrs. Hubbling who told me that she lived in the most exciting time in history. She was wrong. I'm living in the most exciting time—and it can only get better.

Max Pringle OAM is a retired farmer who lives in the small inland town of Narrabri in northern NSW. He is an active volunteer with many community groups, including the Scouts, Narrabri Show Committee, and the local museum. He is currently president of Narrabri Red Cross, on the board of 2MAX FM radio station, a Patron of Narrabri Show, and secretary of the Narrabri Historical Society. In 2008 he was awarded an OAM for services to his local community.

Sarah Pugh

Rising Above

Acerbic voices, punctuated by heavy footsteps on the cracking lino, echoed through the fibro cottage. She shuddered as the chafing sounds found the gap under the door and seeped into the garden to join the resonating pulse of the cicadas. The bitter words threatened to disarm her again, but she had vowed to keep strong. She was her mother's rock. It would not do to crumble. It would pass, of course, the fighting. It always did. She would ride it out: out of sight, out of earshot, out of mind.

She felt the heat of the taut black fabric, still warm after a day in the sun, prickle against her skin. She loved the way the smallest movement of her body caused the skin of the trampoline to quiver, to acknowledge her presence and welcome her in. It had been her refuge in more ways than one. It had swallowed her angry bouncing when Angelina Morse looked at her in that way—'white trash', she seemed to spit without uttering a word. It had also lulled her to sleep after she'd buried her life-long mate, Bernie, who'd picked one too many tussles with the resident Brown. And it had been a friend during the stifling days of summer when the blanket of heat forced all creatures indoors. Even when its shiny surface was too hot to touch, just knowing it was there waiting for the relief of nightfall gave her comfort.

Tonight, it comforted her still, as she gazed upward at the expansive night. In the sky, she forgot her mother's boyfriend, their debts, her longing to join in with the other girls at school, and lost herself in the indelible tales of gods and goddesses, their triumphs and defeats. In this November sky,[6] she knew she would find her Pegasus high in the North. She liked to believe that he was born of the blood of Medusa, that a shining light could come

[6] Information about the November night sky obtained from https://maas. museum/observations/2018/11/01/november-2018-night-sky-audio-guide-transcript-and-sky-chart/

from a creature of darkest intent. This winged-horse, with its grace and power, was her guide in life, and even though most amateur stargazers could also find him with ease, she liked to believe that he drew her to him, even waited for her. She felt her knotted muscles relax as she found him, his strong neck stretching westward. She had never owned a telescope nor visited an observatory. Instead, she explored the sky in yellowing books at her local library. Once she'd memorised the constellations, she was able to imagine the finer details, to create in her mind what her eyes could not see. She wondered about Pegasus's enigmatic hooves. Did springs really burst forth wherever he stamped upon the earth? Were the Nine Muses really born from such springs on Mount Helicon? Was it he who transported Perseus in his quest to save Andromeda?

Andromeda. She knew about this ancient princess from stories she'd read, but what truly beguiled her was the precision of the human eye. Andromeda's Galaxy, at 2.5 light-years from Earth, was one of the most distant objects visible without a telescope. It revealed itself as a long patch of light near her guide's hind leg but all her attempts to mark Andromeda's splendour had failed. She could see the page in the library book in her mind's eye, she knew, from astronomical photographs, every swirl of the dust lanes around the galaxy's bright core, but she wanted to see that smudge of light with her own eyes. She strained every muscle in her body as she willed her eyes to see it. Exhaling—for her lungs had been immobilised by her anticipation—she felt the black fabric beneath her absorb her disappointment. Again.

She lay back, closed her eyes and listened. It was quiet inside the house. The storm was over. When she opened her eyes again, she noticed that the hard angles of the verandah had softened in the shadows of the night. The spider-web of cracks in the concrete steps was no longer visible, nor the heat-warped flyscreens held in place with tacks. The odd assortment of gnomes which, by day, were sentinels placed strategically to bring character to the run-down cottage, had become, in the night, obscure bumps on the grass like mole-holes. In the darkness, her house blended with the others on the street. She relished the anonymity of night.

Returning her gaze upward, she thought of herself as Gaea, the Earth Goddess. She felt alive as if she were the heart of the land with it pulsing rhythmically around her to the incessant thrum of insects. She, like Gaea, often feared the inclinations of men. She conjured in her mind the battle between Orion and the Scorpion. In the East, she saw the shimmering shape of Orion the Hunter. Many knew him as the Saucepan, a name which always brought a wry smile to her face: how ironic that one so famed for his hunting skill and narcissistic boasts that he could kill all life on Earth should be reduced to a symbol of domesticity that was hidden away behind cupboard doors. She thought of him, mortally wounded by her Scorpion, brought down by the very Nature he desired to dominate. Yet, as lore would have it, he would be perpetually revived to return to the hunt.

Like Gaea, she traced this eternal battle across the sky, finding her Scorpion in the West, defeated but destined to be healed and to return to challenge Orion again. She thought of the triflings of man; how their whims and desires all too often came at a high cost, leaving the women of this world to repair the damage. That is what she understood of herself. That she was an observer, a note-taker and compiler. One day soon, if history repeats, she too would have her own cottage, her own block of land, her own brood to mother. But, if grace could intervene, perhaps things could be different for her.

Finding Orion again, she thought of the Pleiades, or the Seven Sisters. She could see six stars tonight, yet she had seen thirteen at times when the town lights were dimmed. She had read the myths about Atlas and his seven daughters numerous times. But she preferred to imagine the Aboriginal Dreaming story of love, loyalty and law. A story that has been recorded in paintings and retold to children for thousands of years. An old Auntie shared it with her class at school, and it lingered in her memory. Each night, low to the Earth, the seven sisters gracefully launch themselves into the night sky to seek refuge from a man of the wrong skin who wants one of them for a wife, a desire forbidden by their complex law and tradition. Each night, he chases them across the sky.

She admired the sisters. Bonded by blood, they sacrificed themselves, in the old Aunty's story, to obey tradition and protect their kin. Such loving loyalty, she thought, was surely a noble sentiment. She thought of Angelina Norse and her 'sisters-in-arms' in the schoolyard and then marvelled at the power of a back, the very structure that keeps a body straight and mobile, which can also be used as a weapon. To turn your back on someone, to conceal your face, to block out, to deny entry: though subtle and often unnoticed by the casual observer, the back can deliver a blow that stings long after the bell to return to class.

She felt the strength in her own spine, now, as she hovered above the ground on the trampoline. Each curve of her body seemed to have melted into the fabric. Above her, she sensed the eyes of the universe on her and relaxed. With the memory of the school bell came the thought of another bell: Jocelyn Bell. She corrected herself: Dame Susan Jocelyn Bell Burnell. The free local had rag published a story about her this year. No front page celebration, though, a mere snippet of recognition on page 7. The word 'benevolent' was used. She had to look it up. Alongside her Pegasus, Bell had now become a beacon in her life. She had scoured the internet at the library, devoured every fact, memorised every detail. What captured her attention was the way in which Bell had shown such grace when her colleagues received a Nobel Prize in Physics in 1974 for a joint project, and she was excluded because she was a graduate student. In 2018, her loyalty and dedication was recognised when she was awarded the three million dollar Breakthrough Prize for her contribution to Fundamental Physics. In an act of benevolence, so the snippet noted, she donated all her winnings to help underrepresented groups gain access to careers in her field. As she lay on the trampoline, her heart swelled to think of Jocelyn Bell, one so selfless who was yet driven to push the boundaries.

Her mother told her once that, when she looked into her eyes for the first time in the delivery room, she saw great wisdom and grace. Gazing at the splendour above her now, she made her usual prayer to the Heavens, to her Pegasus and her Bell, to the imagined past and the untold future, that she may rise up like others before her and be worthy of the name her mother had given her: Grace.

Sarah Pugh is a mother of two who dabbles in writing in her spare time. She lives in Wollongong. An English teacher of 20 years, she decided to put her money where her mouth is and submit some of her work for others to read.

David P. Reiter

Minerva Series

Minerva: to Earthlings

You depict me as woman in the Capitoline
but I am neither – and more powerful:
a deity of voice, freed by an intelligence
beyond decay, embedded in the nano
vacuum that siphons the breath of Jupiter.

Now, as I wait for the countdown,
I am your hope. I dare to gaze down
through the question marks that mist
a trio of booster rockets that will launch
me to the nearest exoplanet and its wobbling

star. My mission will over-dub the marble
of pantheon, the atmosphere of speculation
that nudges brown dwarfs out of orbit.
This is no simulation: I am to dock
with the molten rains of Bellerophon

on its dark side, of course – I'm a goddess,
not a martyr for hot Jupiters, and not about
to be tidally locked at 20000 C for scientists
long since dead to argue their theories.
Water and ozone are the atoms of poetry.

Minerva: Next Contact

My spider-cam has found intelligent
life is a more difficult question
than an answer

I launch a swarm
of curious robots
to whisper in alien

algorithms that test
Drake's Equation of deliberate
fractions for the right planet

at the right time, burbling
in signatures of gold dust
gas and sodium. Do they

recall that Cambrian Explosion
that made predation an art
and evolution our paintbrush?

Should we scramble our radio
signals to camouflage false icons
like the Gorgon or emerging

extinctions? My robots bleep
'no'. We are here to discover
not to be disrobed.

Minerva: Second Genesis

We are ready to explore, decode
solar systems from negative dust
deploy our spider-bots and drones

in our divining for maverick water,
carbon, understudy molecules.
As we biomimic the voice between

silences, not distracted by ice-
whispers from Comet 67P,
Rosetta arranges my samples

into sentient channels that zigzag
around black vacuums that
threaten with reckless energy.

It all boils down to puddles,
waveforms of future portals
shapeshifting before our eyes.

Minerva: Wormholes

There's this part in metastasis
when you throw a grenade in
and the wormhole turns orange

but then retreats back to green –
so what do you do?

Throw in two grenades
that'll destroy it …

Sorry, I was distracted
by chatter about some trailer
I've never played.

I really want to tell you about
black holes as timelords
and the ultimate gasp

of holding on for dear death.
Here, or wherever we are when
there is here, space acts like a

fabric, watch it twist until
it can be measured, warping
space, until it loops into the past

entangling photons into a future
surging faster than the speed of light.
Blue-sky thinking is the best when

we're far apart – or docking in a
fresh parallel. After all, what
we know about past and future

is up/down to math, or the debates
we play with quantum physics
anointing them as new thrones.

David P. Reiter is an award-winning text and digital artist, and CEO at IP (Interactive Publications Pty Ltd) in Brisbane, Australia. His medical/micro-textual hybrid <u>Timelord Dreaming</u> won the 2016 Western Australian Premier's Award for Digital Narrative. As artist-in-residence at the Banff Centre (Canada), he completed <u>My Planets Reunion Memoir Project,</u> which won the 2012 WA Premier's Award <<u>http://ipoz.biz/myplanets</u>> His work-in-progress is A Brief History of Time Lords. <u>google.com/+DrDavidReiter</u>

Alison Rumps

Farewell Neil: words spoken and unspoken

The gathering in the Bowen Room at the Parkes Radio Telescope was elegant. Black dinner-suits wandered past satin-clad beauties, holding their drinks like precious jewels, catching an eye here and there, small smiles, and hello nods. They turned to face the speakers, an older, formidable-looking woman in black, with short white hair, pearls and white-rimmed glasses and, to her left, a grey-haired man, bearded and casual. He held the microphone and introduced her—even though everyone knew her. She was Dr Selene Thomas, for heaven's sake! A CSIRO legend in Parkes, and instrumental in setting up numerous radio astronomy projects over the years.

"Welcome everyone. My name is Paul Donellan and I'll be MC tonight. My wife, Selene, and I are happy to see you all here to farewell our son, my young mate, Neil. We are honoured to have so many of you here, and especially the head of the Australian Space Agency, Dr Megan Clark, over from Adelaide. Thank you all for coming.

"Looking at Neil tonight, it's hard to imagine he's the same boy I used to take camping along the Lachlan River, trekking in Nepal, bungee-jumping in New Zealand. But now he's on an adventure that I can't participate in. He's heading off to the Moon as head of NASA's next lunar landing on the Dark Side. But our first speaker tonight is Neil's proud mother, my beautiful wife, Dr Selene Thomas."

Without acknowledging the quiet clapping, or the knowing smiles that accompanied the word 'beautiful', Selene took the microphone. Emotions that she wasn't prepared for jabbed like fiery darts into her well-armoured heart. Her son's imminent departure had triggered childhood memories of other partings that could never be erased. How was she to make it through this speech when she felt so stretched, she wondered. Not a wisp of emotion showed on her face, however. Selene was steel to her core. Her words were considered even while her thoughts were chaotic.

"Dear honoured guests from the CSIRO and friends and family, welcome to our small gathering. Just let me say a few words before I hand over to my son. All journeys begin with one small step—but Neil will be taking a huge step on this journey.

(Oh God, help me make it through this speech. Small steps can be life-changing. I wouldn't be here today if not for that one small step …)

"We all know the huge part this facility played in the first lunar landing fifty years ago. Neil Armstrong took one small step and made history as the first human to walk on the Moon. July 21, 1969. Do you remember where you were?

"I was fourteen years old and besotted with all things Space— not just science fiction, but the stars, the planets, the galaxies, and how science, engineering and the wonders of pure maths would combine to help us reach out into space."

(School was the highlight of each day and science fiction and the stars were my shields against the harsh reality of life with my father.)

"Times of great importance stick in our minds. Small details are remembered. I was at school in Sydney, Fort Street Girls' High School on Observatory Hill in The Rocks. The teachers lined us all up under the Moreton Bay fig trees at the front of the historic building and wheeled a black and white television onto the marble step of the main building. Neil Armstrong was 384,400 kilometres away, yet we could see him! Amazing in a black and white, fuzzy, static sort of way.

"NASA's Apollo Space Program gave the whole world hope. We finally saw the Earth as it really is, the most beautiful blue and white ball hanging in the darkness of immense space. It was wonderful that Australia, and particularly the Parkes Radio Telescope, could be part of the future NASA was creating."

(The future just had to be better than the present. Home was a misery. No mother. Dead when I was three. Father an embarrassment. The step that was taken by my father could have ruined my life, but here I am. I've blanked him out, erased him.

I've shielded Neil from the shadows of my past, so he'll never be touched by my father's one small step. Oh, he had all his excuses! He'd talk about his time in the Korean War, he'd rail against the Vietnam War, the death, his mates, his loss, his shame, his pain, his anger … everything was about him, him, him. Well, he got rid of his pain with his one small step and left me to pick it up and live with it.)

"That's one small step for [a] man, one giant leap for mankind." Neil Armstrong spoke and a billion people on Earth listened and watched with their own eyes as he took a step off the lunar landing module, The Eagle, and into history."

(I felt so hopeful walking home after the Moon landing that afternoon—but what did I find at home? My father's final legacy. He was hanging, dead, in the kitchen. Small details remain with me: his last cup of coffee, the tilt of his head, the shirt he'd chosen that day, his best Sunday one. My father, overtaken by his shadows, took his one small step from our kitchen stepladder and hung lifeless from the ceiling. July 21, 1969.)

"Neil, we honour the men who have gone before you, men like Neil Armstrong."

(I named you after Armstrong.)

"There are so many who have worked towards a better future for the world—not just the astronauts. The engineers, designers, technicians and administrators, well, I could go on, because so many people have been involved. Even this year, we've seen new steps forward. NASA's New Horizon took photos of that snowman-shaped object Ultima Thule six-point-five billion kilometres from Earth, for example; the Voyager 2 probe officially reached interstellar space; China's Chang'e 4 lunar module landed on the dark side of the Moon …"

(Chang'e—the Chinese Moon goddess. Selene means Moon goddess, too.)

"As for myself, I am at the end of my career and proud of what I've achieved."

(Neil knows my sacrifices. It takes time to be successful. It would mean the world to me if he named his spacecraft after me: the Selene Lunar Landing Module.)

"Soon, we are anticipating the final launch date for your next mission. It might not be Mars for a little while yet, but, Neil, your work is so important in testing our readiness for the Mars expedition. I'm so proud of you, son. I pray for your safety, and I pray that you will always look with hope to the future. Look toward the sun, and as you journey toward it. The light will cast the shadow of any burden behind you, and your vision will be clear and bright."

(You'll experience no shadowy darkness from my father. No bad luck, no sadness, no pain of his will follow you.)

"Now, I'll hand over to Neil. Join me, in raising a glass to Neil, the Captain of NASA's Dark Side of the Moon Lunar Landing Expedition."

Neil took the microphone from his mother, looking every bit the dashing hero—tall, lanky, smiling, casually dressed, feet firmly on the ground, shoulders squared.

"Thank you, Mother. Well, here we are in Parkes, a small town that packs a big punch in the Space business. After all the work, we're almost ready to go, and with just enough time for this quick visit to say goodbye. Let me thank you all for your trust in my ability to see the program through."

(Mother is in a strange mood tonight. She's even more tightly wound than usual. Dad will get her calmed down later, though. Thank God for Dad. He's the one who's kept the family together. He's the one who gave me my love of space. He had the time to sit with me outside at night staring up at the stars and the Moon. He's the one who told me all the names of the galaxies, even some of the Indigenous Australians' stories about the stars. And we spent hours together fiddling with telescopes. Dad's love has made me what I am today. But I wish I could do something to please Mother.)

"As Captain of this expedition, I have the honour of naming our lunar landing module. I had to decide between so many options. I remember the times spent with my Dad out on the oval with the telescope, and the countless hours chatting about the different ways people have understood the stars and sun and Moon. So, I could have named it for him. It would also have been fun to name it for Australia's celestial emu: the Dark Emu Lunar Landing Module! But we all know that the emu has never been known for its ability to fly! And besides, the dark emu is already taken. It's the icon for

CSIRO's Evolutionary Map of the Universe project being done the new Australian Square Kilometre Array Pathfinder radio telescope in Western Australia.

"So, I return to my family for inspiration. We all know it's been fifty years since the first Moon landing, but not many people know that on the very same day that the Eagle landed, Maurice Thomas, my grandfather, died of a heart attack in his home in Sydney. I hope my mother is happy with me honouring her father in this way. His name, Maurice, means 'dark' and, as we are voyaging to the dark side of the Moon, that name seems appropriate. May his legacy bring our team good luck and great happiness in our year-long voyage."

After life as a busy mother and teacher in Sydney, **Alison Rumps** enjoying retirement in her hometown of Grenfell, with its inclusive country community vibe. She is a member of a writers' group, the local church and a fun group of book club ladies. She is proud of her beautiful children and grandchildren, loves reading and her writing investigates the psyche of her characters. She is currently collaborating in producing a history of the businesses on Main Street, Grenfell, called 'Faces in the Street.'

Robert Salt

Birung

My ancestor's gazed upon
(This night *Birung*)
Gazing upon the magnificence of this night *Birung*,
Neck arched,
eyes searching,
mind and emotions unable to process all that is laid before me,
This is part of my essence,
my soul,
my connection to country,
all its natural elements,
A tapestry of constant flux of lights and travelling objects,
The *kiyan*, bright, constant in its scrutiny,

In my sadness I have looked upon its splendour,
In my mother's sadness she has looked upon its slender,
In my father's sadness he has looked upon its spender,
In my happiness I have looked upon its splendour,
In my mother's happiness she has looked upon its slender,
In my father's happiness he has looked upon its spender,
I am linked through culture … genetics … suffering …
dispossession … resilience … sadness,
My ancestors, those who came, those who were here,
They have used the *Birung* for guidance, answers, romance, reflection,
My actions are their actions.
What I see, they seen.
The night *Birung* was and will be there,
when my ashes are across the land

Birung: Wiradjuri Language – Sky
Kyan: Muwararri Language – Moon

Robert Salt was born in Brewarrina and now lives in Dubbo. Culturally, he is connected to the Kunja, Muwawarri and Wiradjuri nations through his mother. On his father's side he is connected to English and French ancestry. He has been an avid reader and writer of all genres and forms since he was six years old. Robert is passionate about Indigenous languages and strives to incorporate both English and First Nations language in his writing. This reflects his duality in identity and respect and gratitude to his mother and father. Robert is also ardent about rural living and the power of words for social change and also enjoyment.

Phil Sanders

Ellipse

For nine years it's journeyed around the Sun in an orbit that falls ever more behind that of Earth, its cameras searching the universe for evidence of other Earths orbiting other suns. Its movements are tracked, instructions sent, by giant radio dishes such as the one nestled in the dry hills of the Australian Capital Territory at Tidbinbilla. From its raw data, light curves have been constructed, brightness values adjusted, conclusions drawn. In its time it's seen sixty-one supernovae, observed half a million stars, discovered close to three thousand planets but now, as November 2018 approaches, it's 94 million miles from home and dying …

He stood in the doorway, framed against the dull November light, leather saddlebag slung over his shoulder. Even in his thick riding coat, he looked as thin as a poplar. The inn, dark as an antechamber of Hell, stank of flat ale, old food, stale sweat and other liquids, bodily in nature, that he didn't care to contemplate. Smoke swirled around the room in the draft from the open door, curling up from a fire of green wood. He'd seen cleaner straw on the floor of a stable. It was no wonder he was gripped with a fever, sleeping in such godless places, breathing in such noxious fumes night after night. Still, if he managed to obtain the 12000 florins that the Emperor owed him he'd be able to stay in more salubrious places on his return journey.

"Shut the door, would you, Your Reverence, that wind's blowing straight from Satan's arse." Cackling laughter echoed from a shadowed alcove. Barbarians! A young girl with a red, round face, twisting a cloth between grubby fists, stepped out of the murk.

"Sorry, sir, sorry. They likes their little jokes, they does." The girl gave a half-bow, a half-curtsey. "Is it a room you're after?"

He nodded wearily as he kicked the door shut with his heel. She led him to the alcove where the drinkers were sitting and addressed one of the company, a thin-faced man with the red and sunken face of a toper. "This gentleman wants a room, father."

"Then you've come to the right place, Your Reverence. Finest inn in Regensburg, eh, lads?"

The company, none of whom looked any more temperate than the landlord, nodded and grinned.

"I'll be here for three or four days. I have business at the Imperial Court."

"Of course you do, sir, of course you do. We gets lots of fine gentlemen staying here who has business with the Emperor."

The irony caused much laughter but the thin gentleman merely gave a slight shake of the head before handing over some coins and following the girl to the rear of the inn and up steep stairs that creaked and sloped away from the wall. The upstairs corridor was as narrow as a coffin. She lifted a latch and stood aside to let him into a room with barely enough space for the bed, with its sagging straw mattress, and a rough wooden chest. "Will you be wanting food?"

"In an hour or so. I need to rest now."

As the daylight faded away, the wind began to pick up, to sweep down from the mountains and moan its way through the streets and alleyways of the town. Making her way towards the thin gentleman's room with a plate of bread and cheese and a flagon of thin wine, the girl glanced out of the small, cracked window on the landing and saw black clouds racing across the face of the Moon, below which a single bright star twinkled. It's very likely that more than an hour had passed, but the inn had been busy and her father was in his cups again.

She knocked on the door. "I have your food, sir."

She knocked again when no reply came and cautiously lifted the latch. It was as well to be wary of gentleman travellers, even the ones who looked like they could be wearing holy vestments beneath their travelling cloaks. He was sitting on the edge of the bed. His face, in the flickering candlelight, was as rutted as a country lane, as yellow as winter corn. He was looking at a leather-bound book laid out on the chest as if it were trying to spy a fish in a deep and murky pond. In his hand, he held a quill paused above an ink-pot.

"Sir?"

"What? Oh, yes, come in. Food, yes."

"There's hot stew downstairs if you wants some."

"No, this will be sufficient."

As she put the platter down on the chest, she stared at the book. She'd not seen many books before. There was the great black bible in the church from which the priest read out God's words and the books that travellers occasionally brought with them and read by the light of the inn's inadequate fire. It was a sublime mystery to her how these dark scratches could become words. The thin gentleman's book also had drawings that looked like cartwheels with lots and lots of spokes. And numbers, row upon row of them, up and down the page. She knew some numbers but these …

"I'm casting a horoscope."

She straightened up, blushing.

"I'm sorry, sir."

"I could cast yours for you if you'd like. When were you born?"

"June, sir. June the sixth."

"At what time?"

"Time?"

"Accurate predictions require precise figures to inform them. I myself was born on the afternoon of 27 December 1571 at 2.30 precisely after being in my mother's womb for 224 days, 9 hours and 53 minutes, which means, therefore, that I was conceived on 16 May 1571 at 4.37 a.m. My father was always an early riser."

She nodded and blushed at this intelligence. Can people really know such things?

"And, do you know, the planets are lined up in almost the same position as they were the day I was born. What do you think to that, eh?"

"I don't rightly know, sir."

"I think, I fear rather, that it is an omen. Run along now."

She curtseyed, noticing as she did so, that there was sweat on the thin gentleman's face, despite the cold.

She lay in her bed, the blanket drawn up to her chin, listening to the keening wind, imagining wolves prowling the streets, red-eyed, looking for prey. Could their guest with the books really look into the future? She'd never had her horoscope drawn up before, not properly. There were old women at the markets who said they could tell your fortune by looking at your palm but this she doubted as they usually turned up at the back door of the inn at nightfall spending their pennies on cheap wine or gut-rot spirits. Would God really entrust such a gift to a ragged bunch of crones who couldn't see their way home most nights? The thin gentleman, though, well, you could tell he was educated, clever. And there were things she wanted to know. First off, would she ever get away from this beer-soaked, tumbledown wreck of a place, away from her useless, drunken sot of a father, maybe, God willing, marry someone who'd treat her right? Would the thin gentleman at least attempt a horoscope for her, even if she didn't know the exact time of her birth?

The morning was as cold and grey as her imagined wolves.

She watched through a frost-rimed window as the thin gentleman rode off astride his slump-backed horse, which wheezed like broken bellows. In the afternoon, she was emptying slops into the gutter when she saw him return walking, dragging his feet along the wet cobbles, his back bent as if he carried a sack of stones.

"What happened to your horse, sir?"

"Sold to a dealer for a good deal less than I paid for him," he said, ribs heaving, his breath catching in his throat. "I fear I was taken advantage of on both occasions. Still, I soon hope to have sufficient funds to purchase a superior beast."

Swaying slightly, he reached out a hand to steady himself against the window sill of the inn. His face was as blotchy as a suet pudding, purple as beetroot.

"Are you all right, sir?"

"I am, alas, prey to many chronic ailments. They come, they go." He straightened up. "Now, excuse me, I have much work to do." She watched him make his weary way into the inn, wondering if she would ever get the chance to ask about the horoscope. She had a few pennies saved, squirrelled away behind a loose brick in the scullery so she could even offer payment.

The pennies tucked into her skirt, a flask of Rhenish wine in her hand, she hurried away from the crowded bar room and up the stairs to the thin gentleman's room. Once again, she knocked and once again there was no reply. She lifted the latch. "It's Susanna, sir. I've brought you some wine."

The candle flickered on the chest, but he wasn't working at his books. Instead, he lay on the cot, his cloak drawn about him, shivering. "Susanna? You're here?" he said weakly.

"Yes, sir. I've brought you some wine."

Forcing himself up on an elbow, he looked directly at her, squinting through the gloom. "Ah, yes, I'm sorry, I thought you were …"

He fell back, racked with a coughing spasm. "My daughter. I have a daughter Susanna."

Despite the shivering, his face dripped with sweat. "You're not well, sir. Shall I fetch a doctor?"

"No, no. As I said, these things come and go. It's nought but fever pustules. Pour me some wine, would you?" He pointed a bony finger at the drinking cup on the chest. She watched him drink then pull the cloak even more firmly about himself.

"I was wondering …"

"Yes?"

"If you were minded to do me a horoscope. My father recollects that I was born just as the sun was rising. I can pay you." She held out the handful of dull coins.

"Put your money away, girl. I am not yet that impoverished."

"I meant no offence, sir." Her hand curled around the pennies.

"I was paid twenty florins for my first horoscope, you know, money I badly needed. And it was a great success. I foretold that the winter would be harsh, that the Turks would attack early in the New Year. Pah! Any peasant worth his salt could have predicted the snows that fell that year. And as for the Turks, well, they weren't gathering on our border to celebrate Christmas, were they? Yes, I could take your pennies, but I'd only tell you what you want to hear. That you'll meet a handsome young man with a small fortune who'll whisk you away from this life of drudgery, give you ten healthy children, and you'll live till you're ninety."

She bit her lip, stared at the floor. He must think her an ignorant fool. Which, sometimes, she thought she was.

"It's not that our futures are not written in the heavens, it's only that we cannot yet read the language they are written in. I've strived over the years to find some method by which I might be privy to God's word but, alas, I have been too diverted by other calls on my time. One has a living to make."

"I'll leave the wine, sir. Will you be wanting food?"

"My mother was an innkeeper's daughter, you know. My grandfather used to stand me on a table, and I'd do calculations for the amusement of the travellers and drinkers. Sometimes I think that's more or less what I've continued doing all my life, perform mathematics for a penny here, a florin there. No, no food. I must rest, continue my work in the morning."

In the morning when Jacob Fischer, the Lutheran pastor, called to see him she led him up the stairs. They found him delirious, lying in a pool of odorous sweat, pointing a finger first at his head and then at the ceiling. He's pointing at the stars, she thinks. The Pastor, cloaked and furred, big as a bear, ordered her to sit by his bedside, which she did, pressing wet towels onto his brow.

"You carry on, my girl," her father said in deference to the Man of God. And then, when he isn't listening, his knees bent in prayer, he whispered: "We don't want him dying before he's paid his bills now, do we? And if he does happen to meet his maker under our roof, we want his friends in high places to know we took good care of him, don't we?"

A doctor was called for, a cheery, red-faced fellow who immediately produced a thumb lancet from a tortoiseshell case. She was made to hold the cup while the doctor, humming a merry tune, made an incision in the crook of the thin gentleman's elbow. "Not squeamish are you, girl?"

She shook her head. She's held a bowl when pigs have been bled, eaten the resulting black pudding. She's mopped the blood from the floor of the inn when words and oaths have turned into fists and blows. She's even taken mass and drunk the thin wine that turns into the blood of Our Saviour, though it's never turned thick and salty in her mouth like her own blood when she sucks on a cut finger.

She sat with him through the afternoon after the priest and the physician left. A broken shutter rattled and creaked in the freshening wind; a cold rain slashed against the roof. He mumbled, he hummed, he played with the ends of the bedsheets. She could not make sense of most of what he said. There's talk of dwarves and comets, of new stars, of witches and dead children.

"Mars … Mars the key, always the key," he said clearly at one point. "Mars, she knows is the red star. Venus is the Morning, sometimes the Evening, Star. Jupiter is the bright one."

As the light faded, he opened his eyes, raised his head, the fever seemingly abated. "Can you hear it? The music. Can you hear it?" There was only the banging shutter and the patter of the rain and iron-hooped cartwheels rattling over the cobbles. There was not even a drunken fiddle being played in the bar room.

"Yes, sir," she said. "I can hear the music."

"It was there all the time, the music of the spheres." He smiled, his face as waxy as a Michaelmas candle. "They say that swans only sing when they're dying. Perhaps we can only hear the music of heaven when we ourselves are dying."

"You're not dying, sir. The doctor said …"

"I'm ready, my dear, quite ready. I've done some good in this life, I think. Not always the best husband or father or son, but I've done little harm. If I had, would God have let me discover the motions of the planets, let me in on the secret I share freely with the world? Why me, why now, I wonder? Why did He not whisper in Ptolemy's ear, save us all from those wheels within wheels and more wheels within wheels?"

"Is it true, sir, as some do say, that our world moves around the Sun?"

"You've heard of Copernicus, then? Is his work much discussed in the tavern of an evening?"

She shook her head, although she thought she had heard the name.

"But we don't, as the good Canon thought, whizz about in a circle like a ball on a piece of string. No, no, that is the subtlety of the Lord. Not a circle but an ellipse."

"An ellipse?"

"A circle, as someone said, that has lost its focus." He laughed or tried to laugh. What came out was a long, drawn-out exhalation that had him gasping for breath, sucking in air as though he were a lost sailor swirling about in a maelstrom. At last, he calmed and lay still, staring at the ceiling, once again pointing through it to the stars.

There was much coming and going of clergymen the next day, 15 November 1630, but it was young Father Donavarus who was with him when his soul passed over. Several old women, black-clad widows, came to lay him out, and the following morning he was borne in procession to the cemetery of St Peter outside the city walls. She followed at a distance and watched from the hillside as the coffin was lowered into the icy grave, unsure of why she was so interested in the dead man. He'd died far from home, and the mourners were few; black-clad clergy and a couple of richly cloaked courtiers representing the Emperor, so she'd heard. As the funeral party hurried back into the city in search of warmth, she watched the gravediggers shovel the earth back into the ground, heard it rattle like hail on the lid of the coffin.

One evening in summer, she slipped away from the inn and met up with the young schoolteacher who waited for her in the shadows by the city gate. The day was still hot, the sky still blue and filled with the hum of insects, the songs of birds. As they walked, he twisted small flowers into a bracelet for her and told her of his day drumming knowledge into the reluctant heads of the sons of the town's burghers. She, in turn, told him stories of the inn, which made him laugh. They took a shortcut through the cemetery. Amongst the blackened, weather dulled tombs, the thin gentleman's gravestone still looked new and smooth although there was no one to care for it, and it would soon go the way of the others.

"What does it say?" She pointed to the deeply chiselled inscription.

"I measured the skies," the young schoolmaster declaimed. "Now the shadows I measure; Skybound was the mind, earthbound the body rests."

They walked on towards the woods, and she slipped her hand into his. "Will you teach me to read and write?" she asked.

"Of course."

"And an ellipse. You must tell me what an ellipse is."

15 November 2018: Tidbinbilla talks to the Kepler Spacecraft for the last time, says farewell and thank for all the data. Safety modes are shut down, and communications severed leaving the craft to drift along on its safe, and elliptical, orbit around the sun.

Fried Snails, Coffee Granules and Stargazey Pie

Sunday afternoon and the rain fell steadily across the Central Tablelands. It seemed set in for the rest of the day, so they sat on the sofa, drinking wine and watching *Jamaica Inn*, the black and white Hitchcock version.

"Just think," Peter said as a rickety wagon rolled across Bodmin Moor, "if my great, great, grandparents hadn't jumped on a boat I could be there, wandering the moors, chewing a straw and talking like a pirate."

"Same here! Well, no! I wouldn't be talking like a pirate, but if my grandparents hadn't emigrated I could be—I don't know— herding goats up a mountain or treading grapes with my bare feet. Cheers, anyway," she said, as she lifted her glass.

Manos Liontakis sat on his heels in the shadows at the entrance to the gorge. The sun had nearly disappeared, a blood-red slice of watermelon dripping into a dark sea where once-upon-a-legend Icarus had fallen to his death. In the east, the sky darkened, the stars began to pulse. He shook some dried berries out of a small canvas bag into his palm then ate them slowly, one by one, reverently. They could, after all, be his last meal. In a few minutes, he would be gone, scrambling up the side of the mountain and across the jagged peaks, to the rendezvous with the Aussie sergeant major and the crazy one-eyed Englishman. He didn't expect to meet up with any German patrols, but you only stayed alive on Crete in these mad times by being alert to every possibility.

It had been a harsh end to the fishing season, the worst that even old Reuben could remember and he was eighty-five, although that was by his own reckoning. Day after day for weeks on end, the seas had leapt and clawed at the harbour walls, smashed against the rocks, sent white clouds of spume soaring up the cliff faces. The rains had swollen the rivers and streams, and these rivers and streams had escaped their banks to run through the village streets. Gales had sent slates flying from the church roof and crashing into the market square. Sleet had swept off the great, grey Atlantic and across the moors. The fishing boats, sturdy two-masted luggers, strained against their ropes and anchors, some of them crashing into the sea wall, others breaking their tethers completely and drifting wildly like violent drunkards thrashing out at their drinking companions. A mile or two along the shore a schooner, in ballast from Penzance, had foundered in a gale. Eight bodies had been washed up on the sands.

In the dawn light, the fishermen gathered on the quayside, hunched in their oilskins, and looked despairingly out across a sea framed by dark clouds. Although their families were hungry and they themselves felt the gnawing shame of impotence, they knew that only empty nets and a plunging death waited for them out on the water. So they returned to their cold cottages and cleaned their already clean gear and inspected their already well-mended nets and waited for the weather to break.

Manos Liontakis bounded up the boulder-strewn hillside using the peculiar loose-limbed gait he'd learned as a boy from his father and which, much to his surprise, his muscles had remembered. Despite the years in Athens sitting at a desk or looking through the eyepiece of a telescope, he was still a Cretan, a descendant of heroes, even of those who had no more substance than a shadow on a cave wall. With German troops in sight of the Acropolis, he'd returned to Crete to see his parents, both now old and frail, intending to return to the mainland to do whatever he could to help. But then he'd seen the Fallshirmjagers tumbling earthwards, parachutes blooming in the clear May skies, and knew he'd have to stay, to defend his home, his island.

The going became easier, and he could look up at the vault of the sky, at the familiar pin-pricked black cloth through which the stars shone. He knew their distances from Earth, their intrinsic brightness, their periods, which of them were giants and which were other galaxies. And he recognised the constellations, those fingerposts by which astronomers found their way around the night sky, and around which the ancients had woven stories of gods and heroes.

Tom Trewin drained his cup of the last of the rum and shrub as he gazed out of the salt-smacked window. He shouldn't be drinking so much he knew, but he couldn't sleep, couldn't stop thinking about what they were going to do if he couldn't get out to sea in the next few days. No fish meant no money to keep the children from going to bed hungry. Earlier that blustery Sunday, he and the other fishermen and their wives and children had gathered in the grim little church with the water dripping on them from the rafters, but their prayers must have been lost in the wind. Still, what else was there to do except hope and pray?

His eyes were drooping now. The hour was late and the rum and shrub strong. He eased himself out of the creaking chair and, as he did so, he realised that the stars were out and the Moon was laying a golden path across the stilled waters.

The Aussie sergeant major and the mad, one-eyed Englishman were waiting for him at the entrance to the cave, fingers on the triggers of their rifles, just to be sure.

"Glad you could make it, Ptolemy, old chap."

The Aussie sergeant major held out a bottle. "Here you go, this'll put lead in your pencil!"

Manos uncorked the bottle and swigged the rough raki. He had no idea half the time what the Aussie sergeant major was talking about. Lead in your pencil? She'll be apples? And who was Blind Freddy?

"Right," said the mad, one-eyed Englishman, "we've got an hour before we head down to the beach. So what's it to be? A sing-song or sit down on the ground and tell sad stories of the deaths of kings?"

Tom Trewin, head now as clear as the night sky, hurried down to the harbour, knocking on doors as he went. "Come along me boys, the fish are a-waiting." They set their nets and gear by the light of the Moon and dancing kerosene lamps and moved cautiously out of the harbour. The wind filled their sails and drove them towards the fishing grounds, across waves that gently lifted and settled the bows of their boats. Maybe there was something to be said for praying after all. As the lights of the harbour disappeared around the hook of the cliffs, Tom looked up at the sky, black as coal tar now the Moon was falling, to where Orion, that mighty hunter, chased the Seven Sisters across the heavens. Six bright Sisters and one shy one whom only the keen-eyed could see. How long has it been, he wondered, since men first looked up at the sky and wove stories out of the stars? He recalled the Reverend Tregorran reading from the Book of Job. "Canst thou bind the sweet influence of the Pleiades or loosen the cords of Orion?" Two thousand years ago, or near enough, and probably more.

They sat at the entrance to the cave, looking out into the darkness.

"That's Orion, ain't it, Prof?" said the Aussie sergeant major, pointing.

"Yes. And those, see there, are the Pleiades, the Seven Sisters. Do you know the story of Orion and the Pleiades?"

Their father held the world on his shoulders, and their mother calmed the winds and the seas and protected sailors from the wrath of Poseidon. Maia was the most beautiful of the Sisters but lived in a cave in the mountains far away from the gaze of men; Alcyone watched over the Wine Dark Sea in the manner of her mother; Asterope bore a child to the God of War, a son in whose honour the games at Olympus would be held; Celaeno was the darkest of the Sisters, having been burned by lightning and counted the Titans, Prometheus and Poseidon, amongst her lovers; Taygete was pursued by Zeus and wooed by Hercules although she avoided the embraces of them both; Electra's son would found the great and unhappy city of Troy; Merope married a mortal for love, a king forever condemned to roll a stone upwards to the very edge of heaven only to have to watch it roll back down. They rarely met, but one day, the day of a great feast, they came together in a forest glade—where they were seen by that mighty hunter, Orion.

As tall as two men, dressed in the hides of bloodied beasts, the hunter was relentless in his pursuit. Luckily for them, the Sisters were blessed with swiftness-of-foot and ran and leapt and turned aside, this way and that, like the deer of Artemis. First, he chased one and then the other, laughing as he did so, for is it not the pursuit that the true hunter loves more than the kill? Perhaps the Sisters should have been flattered to be desired and chased by such a hero, but none of them wished to lie with a man whose beard was like a thorn bush and whose rank smell surrounded him like a cloud of midden-flies. On they ran, through the hot afternoon and into the cool of the evening until Electra caught her foot in a rabbit hole, stumbled and fell. As her sisters stopped their running and looked back, Orion bent over her, panting and grinning, face as red as a satyr, his sweat falling like drops of rain onto her skirts.

Zeus, high on Mount Olympus, had watched the chase throughout the day and marvelled at how fleet-of-foot the Sisters were, how lithe and nimble. But now one of them had been cornered by that oaf, Orion. A thunderbolt seemed to be called for. He looked up at the evening sky where the brightest of the stars were beginning to shine and, with the sort of inspiration that set him apart from the other Olympians, thought of a much better idea …

Orion leaned over the fallen Electra ready to sling her over his shoulders and carry her off to his lair, another conquest, another sporting trophy. But as he did so, as his big, hairy hand was about to grasp her slim white wrist, she began to rise into the air, move past his widening eyes, soar above his head. Bewildered, the hunter looked about him and saw that the other sisters were also flying up into the sky, flying and glowing and growing smaller until they were only points of light.

"Still want to chase them, do you?" a voice boomed from somewhere out of the gloom. "I'll see what I can do!"

The story over, they hefted their rifles onto their shoulders and prepared to move out, to meet up with the party from the British submarine.

"You should come to Cambridge when this lot's over," said the mad one-eyed Englishman who, in calmer times, taught the Classics. "Boffins there with brains the size of dinosaur eggs. You'll be right at home."

"Nah," said the Aussie sergeant major. "Australia's where you want to go. Skies as black as your father's hat and ice-cold beer. What more could you want, eh?"

As the grey dawn seeped over the horizon bringing with it another day of dark clouds, they hauled their nets for the last time and turned the bows of their boats towards home. Exhausted but more at peace than he had been for weeks, Tom Trewin watched the shore come closer and knew that this wasn't the life he wanted for his children, his boy out on the sea at the mercy of the waves, his daughter married to a fisherman wondering every day if he'd come home to her.

After the madness of Crete came the madness of civil war. Communists and government forces fought on Observatory Hill, bullets dented the telescopes, an assistant was killed. Manos Liontakis, now with a wife, and a child on the way, looked southwards.

As they steamed further south the days grew longer and the nights stayed warm. Tom Trewin walked the *SS Great Britain*'s decks, still in his shirtsleeves at midnight, and marvelled once more at the size of her. There were more people stowed away in the decks below his feet than there were in the whole of his hometown, Polperro. Among the sleepers were his wife and children and his two brothers and their families. Ahead of them, still far to the south, was the country where their future lay, while far behind them, half a world away, was the land of their birth and the comrades and kin they'd likely never see again. Not for the first time, he wondered if they'd made the right decision, remembered Kerenza and the children's tears as the wagon containing their possessions had trundled up the hill towards the Bristol road. Whatever happened, life was never going to be the same for any of them.

Even the stars were different. Orion was still there, but he was toppling over. By the time they reached Australia, he'd be on his head.

She spent the afternoon baking in the kitchen from which her husband had been banished. It was to be a surprise. It might also, she thought, be a shock. After watching Jamaica Inn, she had looked up traditional Cornish recipes, which had seemed like a good idea at the time. Now, as she looked again at the illustration on the iPad propped up on her work surface, she wondered if beans on toast ("done to an authentic 19th century Penzance recipe, honest!") might be a better bet.

First, the piecrust: she whisked together flour, salt and mustard and put it to chill. She crisped the bacon and set it aside then prepared the pie filling—stock, crème fraiche, mustard, parsley, lemon juice and hard-boiled eggs.

Next, she skinned and boned the sardines except for their heads, becoming more and more convinced that this pie could spell the end of her marriage! Finally came the assembly. She put the dough into the baking dish and laid the fish out like the quarter hours of a clock with their heads over the side of the dish. In went the filling, on went the piecrust, into the oven went the whole damn thing. She set the timer and poured a large glass of sauvignon blanc.

The sun was setting as she carried the pie, covered with a tea towel, out to the deck. It was October, early spring, and the evenings were beginning to retain the day's warmth. Taking a deep breath, she revealed her creation with a flourish and a dramatic "Ta-Da". The piecrust was golden brown and the fish heads dark and crisp, their dull, dead eyes staring heavenwards.

"Oh, my god, it's stargazey pie! You know, it's the reason my ancestors emigrated."

"What?"

"Joking. It's beautiful. Just like the cook. I haven't had this since I was about ten. At my grandparents' house. They told me some old story about a fisherman saving his village from starvation by going out in a storm and catching seven types of fish, which they baked in a pie. Fantastic."

She sat down and relaxed. Whew! Next week she'd do something from her Greek ancestry. Maybe fried snails, done the way her grandfather liked them, to his mother's traditional Cretan recipe. Or maybe not! Souvlaki might be the way to go!

The sky grew dark, the air chilly, as they sipped their wine and finished the pie. All they left were the fish heads for the dogs' supper.

"Look," he said, pointing, "a shooting star."

"There's another one. You know they're only tiny little grains of dust, the size of coffee granules."

"Oh, yes? How come they killed off the dinosaurs and destroyed half of Siberia, then?"

"Those were meteorites, completely different. See how they seem to be coming from the same point in the sky?"

"That's Orion, isn't it? You know, the hunter. I remember the story from school. He's chasing the Seven Sisters across the sky."

"And never getting any closer."

"I suppose not, no."

And so they finished their wine and cleared away the dinner things and turned off the lights and left the night to the stars, to the specks of cometary dust burning up in the atmosphere, to Orion in his vain pursuit of the Seven Sisters, to the Sisters themselves who might be daughters of Atlas, who might be the sky-swept women of the Wiradjuri People. They might also be the Orphan Boys of the Blackfoot, the Seven Sages of the Hindus or perhaps the Subaru of the Japanese. And their pursuer might be Sah, the father of the Egyptian Gods or the True Shepherd of the Sumerians.

As the night sky tumbles around the Earth, it takes its stories with it, through space, across time. If only more of us would look up more often …

Phil Sanders used to be a writer but now has a proper job. Born in the UK, he moved to Australia in 1995 to work for Channel 7 as a script writer/script editor on *Home and Away*. He wrote over 200 episodes of the show plus scripts for several animated series such as *Blinky Bill* and *Old Tom*. He was nominated for three Australian Writers' Guild Awards. Phil lives in Bathurst where he is the Research Office Manager for the local health district. He has written two unpublished novels and is optimistically working on a third as well as some radio and theatre projects.

Braam Smit

Tree Dance

Racing out of a quiet black night
on a desolate country road
away from the darkness and away from the cold
I saw at every swerve and turn, suddenly appearing in the beam
of the light,
majestic trees pressing against the edge of the road, with their big arms
raised in theatrical poses
as they danced and acted out secret and ancient tales
that whisper from this wide and open heart
of central New South Wales.

The stars, a packed audience viewing from the galleries above
excitedly spurred them on with silent chants
that would repeatedly build up to a mighty celestial roar

until in the east, adhering to a kookaburra's call
the sun started brewing a deep pot of molten copper, boiling and
spattering
over the distant brim across the soft blue flesh of a newborn sky.

I continued driving into the morning
still swerving and cutting through the plains
but the trees now stood next to the roadside staring,
far removed from the glamorous night
and stripped from their costumes by daylight
with a hot wind pulling on the dusty brown rags they were wearing,
as they attended to their tedious tree labour throughout the
scorching day

until, in the east, the first stars took their seats,
eagerly calling out for another performance
on this old and rugged crust of sacred planet Earth
to celebrate the dawn of life and all things breathing
and whispers that echo on through time
at the times when they are dreaming.

Braam Smit has always had a passion for writing and wrote poetry from the time he was able to write. His poetry has been published in both South Africa and The Netherlands. His family relocated to Parkes in New South Wales during the winter of 2018. The rugged beauty of the Australian landscape deeply moved him and has inspired his writing ever since. He is also an amateur astronomer and loves spending time behind his telescope. The Skywriters Project, therefore, embraces two of his passions: poetry and astronomy.

Alicia Sometimes

Perspective

i.

From this vantage, Mercury and Mars hang parenthetical
closed sentences while the rest of the galaxy is translucent.
The stars, floating caravels in a mesmerising battalion.

This hill, with its cape of wind and ebb of solace,
allows me to reach out and stroke Jupiter's Moons.
Peering into the beginnings of things.

You stand beside me in that tan, torn coat
as stellar showers squint in the dark face of time.

ii.

How large our curiosity looms.
Your knot-thick hands clasp the creases of volcanic ridges.

These figures eminent, exclamation marks to history.
You said it's important to see more than we're told to.

Peering up from the observatory
on Siding Spring Mountain
deciphering knowledge perpendicular.

iii.

Now, the Moon is almost hidden
Every note from Bob Dylan's lips
falls to the ground perfectly re-formed.
Each vowel running its fingers over my back
anticipation of answers and comfort.

Lyrics, bending chronology.
Orpheus himself weeping.
We talk about absolutely everything.
Hoarding hyperbole.

We are astronomical interferometers
calculating our distance.

iv.

Simone Weil said: 'Truth is on this side of death'.
The cat is both alive and dead and looking out the window.

Warrumbungle National Park cradling all hope.

v.

Astronomers rarely look up at the sky.
Instruments detect invisible signals.

Lists of graphs and diagrams and numbers
chart the unknown and unheard and unsung.

You come home and warm up by the fire.
You open your mouth to say something.

Words, untamed as a strand of string.
Possibilities open up like a box.

Alicia Sometimes is an Australian writer and broadcaster. She has performed her spoken word and poetry at many venues, festivals and events around the world. Her poems have been in *Best Australian Science Writing, Best Australian Poems, Overland, Southerly, Meanjin,* ABC TV's *Sunday Arts* and more. She is one-sixth of the Outer Sanctum podcast (ABC), and is director and co-writer of the science-poetry planetarium shows, *Elemental,* and *Particle/Wave.* Alicia is currently a Science Gallery Melbourne 'Leonardo' (creative advisor) and is passionate about combining science and art.

Tracy Sorensen

Helioseismology

I

I burn carbon to see you
And you burn carbon seeing me

When you glance up from your screens
Your faces glow with the light of ancient forests

Sea-creatures skittered across the sand
Fed themselves somehow
And gave their bodies to you

For this light

I want you to talk to you about my father
Your grandfather

I have made him into pixels for you
I caught his voice in wave-form for you

But some pixels are fading like stars going out
Leaving some areas hard to interpret
And once-smooth lines are jagged now
Not worn smooth the old analogue way

We were shrieking with youth for a little while
Your mother and I
On VHS

But the machine that played us
Got stuck

My voice is a wave-form for you
Green on black, going up and down
Cutting in and out

It's saying something indistinct
Like Armstrong landing on the Moon through snow
Important but not always intelligible

It's all there in the code
The code that says live
The code that says die
The algorithms of connection and disconnection
Companionship and rage

He was a tyre retreader
Cutting new grooves into worn tyres
The old analogue way
The idea was to keep them going as long as possible
Before they got down to the
Fraying bone-coloured string

The machine that played us
Is e-Waste now

Your mother and I are caught mid-shriek
On the magnetic tape
Now stilled across the heads

I beep
With longer silences between one beep and the next
My battery is running low
My telomeres are fraying

I wanted to tell you something
But maybe I don't need to
Maybe it's all there in that softly glowing light

II

Two scientists from the University of Birmingham
Arrive and set up their things
In a tiny domed building
On a red dune

The sun breathes and flares
It runs on an eleven-year cycle of turbulence and calmer times
It is responsible for everything here:
The dune, the scientists, the bundles of wires running up and
down the walls
My eyes reading the graffiti on the door:
Fed the sandflies June '96
Shutter shennanigans Feb 2015

A tiny dot of sunlight enters the dome
It skips through a series of lenses
And into a chamber of potassium burning lilac

A digital heliotrope
Collecting doppler shifts
That tell stories too hot to touch

Actually all I want to say
In my green wave-form voice
Coming up from the bottom of a dying sea
Is this:

Your grandfather was a young man
With black hair
He caught a big fish from a jetty
It was the time of the Moon landing
He was wearing a striped shirt
He looked a little like Victor Mature
In the pub they called him Zorro
He drank Swan Lager from a 7 fluid ounce glass
And once fell in through the fly screen door and lay there on the Lino
Sleeping it off

He had a yellow wax crayon
Just like the ones we had at school
But his was for marking tyres

I still have one with the dirt of his hands on it
Under glass, if you'd like to see it

Tracy Sorensen grew up in Carnarvon on the north coast of Western Australia and now lives in Bathurst. She has worked as a journalist, arts administrator, university tutor, freelance video maker and publicist. Her debut novel, *The Lucky Galah*, published by Picador in 2018, was long-listed for the 2019 Miles Franklin Award. It tells the story of the Moon landing from the point of view of a galah who has just shredded Donald Horne's *The Lucky Country* for exercise. Tracy will be the 2020 Writer in Residence at Sydney University's Charles Perkins Centre.

Stephen Turner

Dark Sky Dreams

TODAY

Dad says I sometimes talk too much, but I think I talk when I need to, and I usually have something good to say, but not when Ben tells me off. Then I just tell him to shut up. But, usually, I've got something in my head that has to get out so I talk and talk and talk and talk until I can't think of anything more to say. I know Dad doesn't really mind, even if he sounds grumpy. Did you know that Acaridians have four arms and four eyes? I'd love to have four arms, but I think the eyes would be hard. How do you keep them clean in the pool, would you have to wear two sets of googles?

My teacher also thinks I talk too much sometimes, too, especially when we're supposed to be doing a test. I do pretty well at tests most of the time, but I don't like them much. Year 4 can be boring. My teacher is nice, but things are too easy.

Coonabarabran is my home and I think it's pretty good here. My house is good and school is close, and you can ride your scooter around town, and we have good Internet so I can play my games. But the best of all are the observatories. The big one at Siding Spring lets you go in once a year, and there's even small ones you can visit any time and even get to use! Going to Siding Spring is like a second Christmas. You don't get presents exactly, but you get to see so much cool stuff!

But the most amazing thing about Coona is the skies. It's hardly rained at all for ages so most nights the skies are clear and the stars are bright. Every night you get to see so much: the clouds of the Milky Way, the Emu and even the Magellanic Clouds that my friend Donna taught me about.

I can just lie in the yard all night in summer, staring at the sky and imagining what it must be like up there. You can look through telescopes if you like, but, for me, just staring into the dark night sky is when my ideas come out. I can see all sorts of things, like where I might be someday …

TOMORROW

"You are aware, Professor, that this operation is under the highest security clearance?" The voice was coarse, and I didn't like the speaker at all. But he was in charge.

I nodded and showed none of my doubt. "Absolutely." A pause. "But why do you want an astrophysicist?"

Barker shuffled through the papers in front of him. My office didn't have a single slip of paper in it, but I'd heard that government types still stuck with the stuff because it was the only way they could be sure their data wouldn't get hacked or stolen.

"You're famous," Barker said. "We need that." He paused, then added, "Your skills might come in handy, too."

I smiled. It was true, even though I wouldn't say so myself. I was an astrophysicist, but more significantly to most people, my channels had over eighteen million followers, and I was featured on both the main HoloBroadcast channel and several sub-channels. I'd taught millions about deep space for most of my life. It was paying the bills, that's for sure!

"I used to watch you on YouTube in its dying days when I was a kid," Barker said. That surprised me; I thought he was older.

"That's nice of you to say. It's nice to meet a fan."

"But it's not just for PR; it's practical," Barker pointed out. "Whatever that thing is, we need communicators. No one's saying it, but we have no idea what we're dealing with."

I glanced out the window. It was there again, as always.

The Thing. Across the centre of the sky, a large perfect isosceles triangle, shiny and metallic. Perfectly silver and utterly featureless. We'd all seen the close-up satellite images before the satellites stopped functioning. Perfect Silver, that's what people were calling it.

"Nothing you've tried has worked, has it?" I asked.

"We've tried every single kind of communication that seems possible," Barker replied. "You know there's been nothing."

I nodded. I'd seen the latest briefing, so none of this surprised me. Everything from DNA frequency pulses to Morse Code had been tried from multiple sites around the world, without result.

"So we're going to do this then?"

It was a multi-national mission. A seven-member crew from the USA, China, Japan, Germany, Sweden, Uganda and little old me from Australia. Four men and three women.

We'd trained together for weeks, always knowing that we didn't have much time. Glover and Wu were already pilots-turned-astronauts, having trained in their national programs their whole adult lives. Satori, Schmidt and Olsen were scientists like me but had some aerospace training, mostly private. Kaluuya and I were the only one without any space experience. His background was in biology. And then there was me. I'd never felt Imposter Syndrome in my life, but if there was any time for it …

"Go for launch, Pytheas," Director Yau announced over the radio. Pytheas was the semi-mythical Greek explorer said to have been the first person from the "civilised" world to explore Northern Europe and chart Britain. That was over 2400 years ago. Pytheas, if he existed, went where no one else in his time and place had travelled, to parts of the world that "civilised people" had never seen. I don't know who had given us his name, but perhaps it was their idea of a joke.

"All systems ready," Glover, our American crew chief, announced.

None of the training could prepare me for the launch. The engine rumbled and then roared to life, blocking out every other sense. The thrust hit me, a physical wave that slapped every muscle in my body, an invisible wall that we had to push through. I felt everything in my stomach getting ready to come straight back out, but I held it in and kept my nerve.

Then we were moving. A sudden, fast rise. We were away! I went immediately from nausea to jubilation, and it took every bit of my training to stop myself from whooping and cheering. Who else would ever get to do this?

The rocket thrust grew stronger, pressing me back even with the pressurised suit. We had no view of outside, but there were several monitors placed to let us see our destination.

The Silver grew closer. It scared me but excited me even more. Any minute now the engines would probably cut, and we'd be dependent on our personal space suits. My whole life had led to this.

Something on The Silver started to glow. An airlock maybe? A place to land?

We were closer now. I could see individual parts of the ship. It was still silver, but there were visible differences up here, different sections maybe. I was ready for it! Whatever *it* was.

TODAY

Until then, I dream, I work, and I wonder.

Dad says I talk too much, but I just want to tell people what I'm thinking. Always thinking. Always about the future.

"Time for bed, Lock," Dad calls.

I roll over and sit up. The blanket and the ground are warm in the late summer night. "Five more minutes?"

"Only five more, mate. It's late."

"Okay, Dad," I lay back to take one last look at the sky.

It's my sky, this little patch that offers so much. It is there for me, so easy, so ready, and every night if I want it. I can imagine my future up there. A future for everybody—if they want it, too.

Stephen Turner grew up in Sydney but now lives in Coonabarabran, the "Astronomy Capital of Australia" where he writes and runs an IT business, as well as doing IT support for the local high school. "Dark Sky Dreams" is certainly fiction, but he doesn't deny that one of his four sons may strongly resemble the main character in the present! While having a past in journalism and being a finalist in several writing competitions, this is his first professionally published fiction. He is passionate about storytelling, technology, the Internet and science fiction. http://stephen-turner.net

Robyn Warrick

Mystic Moon

A full moon is rising over Mount Kaputar, the once-active volcano in the Nandewar Ranges near Narrabri. I can see it from my front yard as it slides effortlessly over the peak to illuminate the town and farmlands beyond. How many times have I watched the Moon rising over Kaputar in my sixty-plus years, I wonder? It must be hundreds by now.

As a child, I moved from one farm to another and then from one town to another with my parents and siblings. With no aunts, uncles, cousins or grandparents, I felt like a lonely little star twinkling in the vastness of the Inland. By the time I was nineteen, I'd lived in seventeen different houses. Perhaps this is why, as I matured into adulthood, I developed such a strong need for a sense of belonging. Even in my twenties, I wanted to know where I came from. I was desperate to find a connection to the past. And so began what has now been forty years of family history research.

My ancestors arrived in Australia from England, Ireland and the west coast of Scotland in the nineteenth century. Some of them probably spoke Celtic languages, including Gaelic. I've corresponded with a distant cousin from my mother's father's line who remembers her parents speaking Gaelic when they didn't want the children to know what they were talking about!

Discovering my ancestors' cultural heritage has made me more curious about Celtic beliefs, folklore and legends. I think of my Celtic forebears often. Especially at full moon, because I now know how important the Moon was to them. A broken bronze tablet excavated in 1897, at Coligny, a small town near the city of Lyon in the Auvergne-Rhône-Alpes region of France, confirms the significance of the Moon to these ancestors. The tablet dates from the 2nd century CE when Celtic Gaul and Britain were both parts of the Roman Empire. When the tablet pieces were re-assembled they formed a lunisolar calendar, which showed the Celtic months.

Each month began with a full moon and included a 'light' and 'dark' period, as determined by the moon's waxing and waning. The Coligny Tablet, as this calendar is now known, is kept in the Gallo-Roman Museum of Lyon-Fourvière.

I shouldn't have been surprised that Celtic and other ancient peoples revered the Moon. It's such a dominant feature of the night sky, so how could it *not* be important to them? In some cultures, the full moon was believed to be a goddess. In some traditions, it has a destructive side and it traveled through the night hunting for victims to consume. Some groups saw the Moon as a "man-eater". Its pock-marked appearance and effects on human behaviour are accounted for in numerous legends. People who act strangely are still said to be "moonstruck", and lunacy, a term for madness, comes from the Latin name for the moon goddess herself. Once, superstitions about the Moon's evil influence made people refuse to sleep in places where moonbeams could touch them. The thirteenth-century English philosopher and priest, Roger Bacon, warned people of this danger: "Many have died from not protecting themselves from the rays of the Moon," he wrote.

My own most memorable experience of the Moon came in 1969 when all the staff of the local newspaper where I worked were summoned downstairs to the printing room to watch, on a tiny black-and-white TV set, humanity's first historic steps on the Moon's surface. We marvelled that humans could actually construct machines to take them into space and land them on the Moon.

I'm still amazed by this technological feat, but, in my more mature years, I'm more attracted to different ways of thinking about the Moon. My late father was very influential in this regard. He was a very keen gardener who sowed and harvested by the Moon. I suspect he was carrying on ancient traditions passed on by his parents and grandparents. Unfortunately, I didn't ask him about this. I just took his moon gardening for granted. I've since learned, however, that his gardening practices were similar to those of biodynamic farmers who harness the energising forces of the Sun, Moon, and stars.

Like my Dad, biodynamic vignetter, Rod Windrim, who has a vineyard in the Hunter Valley, aligns his soil preparation, planting, pruning and harvesting with the lunar cycles and with other natural rhythms, for example. He does this because, as he describes on the vineyard website:[7]

> It's better for the environment, better for the people that live and work on the farm and it gives us a better quality wine that is ultimately better for our consumers to enjoy!

My Dad would have approved!

Rod Windrim and other biodynamic growers see the Earth as a living entity that inhales and exhales according to the phases of the Moon. They claim our planet exhales during the Moon's waning phases when its cosmic energy moves to plant roots, and inhales during its waxing phases when energy moves to leaves and stems. I liken this to the ancient Celtic belief in 'light' and 'dark' periods of the month, as shown on the 1,800-year-old Coligny Tablet.

I love the idea of gardening in time with the earth's 'breathing' in and out, and the parallels between moon gardening and ancient Celtic beliefs. I also relate biodynamic farming to the practices of reflexology, Reiki and crystal healing, which are also about energy flows. There's no scientific evidence that these healing practices 'work' but many people get a lot of relief from them. Scientists might call this a placebo effect, but the healing is real for those who experience it.

I'm the first to acknowledge that my understanding of science is very limited. I'd really like to understand more, though. I recently asked my granddaughter to download an app called *SkyView* to my phone, so I could learn more about the stars and space. I found the constellations fascinating but of even more interest to me was the Hubble Space Telescope and the International Space Station. I'm not sure why. Maybe it's something to do the contrast between these contemporary technologies and the moon landing I watched back in 1969.

But tonight, as I gaze at the Moon barefoot in my garden in Narrabri, I'm thinking of my ancestors who, six generations ago, left their familiar Scottish sky and, after months at sea, disembarked on the other side of our planet. What did they think as they stepped onto this ancient land of ours and looked up into the unfamiliar

---

[7] https://www.krinklewood.com/biodynamics/

southern sky? Even the Moon would have looked different to them. They were farmers, these ancestors of mine. They surely respected nature and, like my Dad, probably planted and reaped their crops by the cycles of the Moon. As I ritually lay my crystal collection on the earth beside me here in Narrabri, I feel I'm continuing their ancient Celtic traditions and sharing their belief in the revitalising power of the universe.

I hold my new app skywards to identify the constellations and planets that are visible tonight and seek out the International Space Station and the Hubble Space Telescope. We know so much more about the universe than my Gaelic-speaking ancestors knew when they arrived in Australia. The contrast between then and now is astonishing—and yet so many of my ancestors' beliefs still course through my veins.

After forty years of searching, I feel a sense of belonging. I feel connected: with my ancestors, with the universe, and with the planet, as represented by Mount Kaputar and the Nandewar Ranges in the distance and here, beneath my feet, by my garden. At last, I'm at peace.

Robyn Warrick lives in Narrabri, in north-western NSW. For several decades, she worked as an advocate for people with disabilities and, during that time, gained a Graduate Diploma in Social Science and Community Services. She turned to creative writing after the tragic loss of not one but two husbands. Her current priorities include her family, connecting with nature, gardening, photography, family history, and personal wellbeing.

Leanne Wicks

Jealousy

The Pleiades laughed
    as we kissed that night.
Yet they said nothing as we told
    ourselves to each other
        among the Dabee shadows.

Sisters! Why did you not warn me that he would be the one
    to save me from finding love?
You only winked when I stared
    at the constellations as he
        filled me with memories.

His farewell embrace … Oh, I'd give my pen
    to taste again his songs,
        to be his purple-winged wren.

Sky Gems, 1872

Bushrangers. Every dusty movement along the Cobb and Co roads hinted at them, each heavy knock on out-of-the-way doors sparked concern. The *Gunnedah News* ran a dedicated column about the latest outrages of these clever cads. Gentlemen would shake their heads, women eavesdrop, and boys widened their eyes with fear and anticipation.

On Wednesday 21st August, William Strong's family were in the kitchen of his stage post, The Coach and Horses Hotel.

"Why can't the police nab this Captain Comet?" William lamented. "Our business will be ruined if he's at large much longer. What if he comes here?"

"Maybe he has already been here, Da," his son Robert said. "We'd never know! Remember how the Murphys seemed to fit in most places before they were captured."

"Apparently, he lets people go if he likes them, Da," his sister Sarah interjected. "Listen to this:

> Mr Smith of Rushes Creek informed *The News* that he
> was lucky to be alive after a run-in with the bushranger
> Captain Comet. On Friday last, Mr Smith was heading for
> town in his buggy to take his wife to the doctor as a matter
> of urgency.
>
> Along Swamp Road, Mr Smith was approached by a
> man on a tall, bay mare. He matched the description of
> Captain Comet: 5' 8" with grey eyes. He was, however,
> wearing a half-kerchief over his nose and mouth. The
> bushranger asked where Mr Smith was going and drew
> a pistol from his holster. Mr Smith relayed his present
> medical urgency and Mrs Smith began to swoon with the
> danger of the situation.
>
> Captain Comet replaced his pistol and released the couple
> with his best wishes to continue their journey, informing
> Mr Smith that his money would be needed for the doctor."

Robert shook his head. "A considerate bushranger? Whatever next?"

"We do have the constables visit regularly—that should keep any felons away," Sister Annie surmised.

"This comet is in Inverell one minute and one hundred miles away the next," said Robert. "How does he travel so fast?"

"I've heard that the Browns in Vegetable Creek swap horses for him when he rides through," answered Sarah.

This eldest of the Miss Strongs was sociable and listened to all-comers about the likelihood of drought, how Earn's leg was playing up or the political state of the colony. She was always receiving compliments and smiles across the bar.

"Back to work, I think," said Annie, returning to the small room behind the bar where she diligently kept the books. The heavy wooden desk was the only piece of furniture they had brought from Strathbogie, an appreciated gift from Hugh Gordon, Esquire, owner of the property. He was a member of New South Wales Parliament and a believer in free education for all.

Annie fondly remembered the mathematics lessons delivered by Mrs Gordon and her tight bun on the back of her head that held as solidly as a dried-up cowpat.

The early afternoon chatter of Willie Wagtails was interrupted by William Strong checking on his daughter's progress with reconciling of the latest delivery of spirits. "Make sure you record each bottle, my girl. We need to stretch every drop."

Annie had been inking in figures with her nibbed pen all morning. "Yes, Da. I think we'll make the bills this month." She had taken over keeping the books since her mother passed.

The grief of leaving Strathbogie weighed heavily on them all since the discovery of tin on the property. Mr Strong had gone in with a mate to buy The Coach and Horses in Gunnedah and had predicted making great profits from gem hunters. "Gems will be the next gold rush," he had said confidently.

His wife, Hannah Strong, convinced herself that their new life would be brilliant. She had safely delivered most of the children in the district and continued to midwife. The locals now called her 'Granny Strong'. Many times on her way to a birth in the wee hours, Hannah would pray for the babies and mothers, looking up at the Milky Way as her horse plodded on. She imagined that the stars would bring light and blessings to her patients.

One December day the previous year Hannah complained of a sore throat and woke up the next morning with a strawberry tongue and sandpaper rash all over her body.

Dr Richmond knew the signs immediately. "Scarlet fever. It's the little ones that usually succumb. Mrs Strong, you must remain confined. It is contagious. I shall visit again in a week's time."

The worst part for the children was keeping their distance. Annie and Sarah took turns taking in meals, although, with the nausea that accompanied this illness, their mother could rarely hold her food down.

Just as Annie put away her pen and ink, she heard the clop and panting of horses.

"Get ready, Sis, we're in for a rush," Robert said. "There's a commotion outside."

Sarah, too, had heard the raucous cheers on the road and emerged from the kitchen into the bar. Annie darted around the side of the pub and peeked down the road. She saw dust clouds and, beneath them, three mounted police troopers with a trail of locals waving their hats, cheering, and chattering like a mob of white-winged choughs. The troopers were slapped on the back as they tied up their horses to the timber railing along the front verandah.

As the stable boy came to see to the horses, Annie stole back to the bar to start pouring drinks. She turned to look into the large mirror on the sidewall, the one with the advertisement for Old Crow Whisky and the date the pub was established: 1831. After checking her appearance, she adjusted her neck scarf, slid her palms across her bustle to smooth it, tucked a stray wisp of brown hair into her braid and silently wondered why she was the only one of the five children *not* to inherit Hannah's dragonfly-red hair.

Her musings were interrupted by three troopers striding into the bar in calf-high, black police boots and their royal blue uniforms, followed by the crowd of supporters. One of the troopers was holding up a Martini-Henry breech-loading single-shot rifle, as if it were Moses' staff.

"Captain Comet has fallen, and this is the gun that did it!" Sergeant James Shannon declared to a fresh wave of backslapping and cheers.

"We got him! We bloody-well got him!" Trooper John Jones announced, rising on his toes and thumbing his jodhpurs. "He won't be passing by here anymore."

The crowd surged like a flash flood towards the bar.

"Give us an Old Crow each, Annie!" glowed the Sergeant.

Reaching up behind her, Annie stood on tip-toes to retrieve the whisky. She filled the two glasses a little more than usual.

"Congratulations, James," she said. "And you, too, John. You are quite the heroes!"

"And one for Trooper Crowley," Sergeant Shannon shouted. "He fired the winning shot. Not bad for your last day in the force, eh, Thomas?"

Annie poured another whiskey, pausing as she placed the glass in front of the newcomer. His wavy hair and dancing beard framed a tanned complexion, determined eyes and a straight nose. Thomas held her gaze and downed the liquor.

Annie found herself scanning Thomas' five-foot-nine frame as she would a tally sheet. He glanced at her. She snapped back to work, focusing on serving the crowd. She was sure that she felt his gaze following her as she collected glasses. She wiped the tables a little slower than usual. Her shoulders seemed to shiver like the small bubbles that form just before water boils in the pot.

"Annie." Robert nudged her as she returned to the bar, "I haven't seen you smile for months. You remind me of Mum."

Annie glanced into the bar mirror and was pleased to see that her brother was right. The afternoon sun flickered through gum leaves and windows and danced across the room to her face where it landed as softly as the Irish snowflakes Hannah so often reminisced about. She shivered under the star-white words etched into the mirror: *Established 1831*—her mother's birth year.

Just before she turned towards the bar, Annie's eyes lingered on the mirror and connected with Thomas' gaze. The reflection was deceptive, however, because, although the bar was between them, they appeared to be close together. He stood as if posing for a photograph. Annie noticed the police number on his collar and drew a sharp breath: 31. Was this a coincidence, or was the universe telling her something?

The forgotten noise of the room jolted her as Sergeant Shannon asked Thomas what he planned on doing now that his policing days were over.

"Sapphire hunting, James," he replied. "A few friends have struck it lucky lately. I reckon that finding a big, beautiful gem will set me up for life. At least Captain Comet and his likes won't be shooting at me!" He glanced across the bar at Annie. "And I might have found a gem already," he said.

Small Reminder

When I was a boy I learned all sorts of things from the blackfellows. They soon got friendly and I was sorry to see them gradually dying out. I had many playmates amongst them and I haven't forgotten the principle words of the tribes round here yet. (Here the old gentleman quoted a score of native names for … sun, moon, stars).

– Centennial Supplement to the Daily Telegraph
Sydney. Monday, January 23, 1888
Mr William Small, born 14 Dec 1796

Sirius led the fleet-footed
Kingsmen south
I am the son of a convict
didn't want to be here,
worker ant for the colony.
My sister the first white child
but not the first bright, wild
daughter of the land.

Dark hands waved
by the river,
we were boys
irreverently
skimming stones,
skin tones irrelevant.
Laughter rolled around
our gums
under the same sky's sun.
Our dream Moon
slipped into wide smiles
in love with the world
right down to bare feet
pressing on the past.

I still gaze up at
slayed playmates
displayed as
stars.

The Sky Tree

The shining ones fell
　　　from the whorl of night
　　　　　and squawked freely.
Five birds now in this world
　　　announcing His return
　　　　　to those who know the Word.
White cockatoos glow in the gumtree,
　　　gravity claimed them from the sky,
　　　　　Crux to Cross.

　　　　　Our Lord's

　　wounds　　　　　　　echo,

　　　　　through

　　　　space

Twinkle, Twinkle

Twinkle, Twinkle Battle Star
Twinkle, twinkle little star
once I wished on you afar:
"Lift me from domestic cries
and spare me from all his lies."
Twinkle, twinkle hopeless star
can you stop this violent war?

Twinkle, twinkle helpless star,
you shoot across like my scar,
you shine like my bruised right eye –
only violence in your sky.
Twinkle, twinkle little star
how I now know what you are.

Leanne Wicks lived in the central-west towns of Kandos, Lue and
Mudgee for over 15 years. She developed museum exhibits, taught
poetry and performed belly dance. Leanne has self-published three books
including *Against the Skyline,* on the sinking of the Australian Hospital
Ship *Centaur*, which one of her relatives survived. In 2018 she won first
prize in the poetry section of the FAWQ Literary Competition. Leanne
recently moved, with her son, to the Victorian seaside town of Mallacoota,
where she now knits birds and writes. Her current projects include
completing a Graduate Diploma of Creative Writing, and a book about the
church where she manages the op shops.

Wing-Fai Wong and Juanita Kwok

Chinese Astrology in Australia: A Case Study of a Chinese Astrologer on the Turon

In November 1865, a correspondent from the Freeman's Journal wrote that a Chinese 'astrologer' on the Turon River goldfields had accurately forecast rain. The astrologer had for months predicted that rain would fall on the first Friday in November and persistently asserted this. At one o'clock on the predicted date, rain fell steadily for twelve hours, and from time to time afterwards.[8] The method used to predict the rain was not described. The astrologer may have used zhanyu 占雨, literally rain divination. Amongst the many different zhanyu techniques are spirit-writing (fuji 扶乩) and the use of wooden divination blocks, jiaobei 筊杯. These techniques are reported to have been used in a forecast of rain made at a Chinese temple in Sydney in 1902.[9] Alternatively, the astrologer may have used Chinese astrology, xingzhan 星占. Up until the end of the Qing dynasty in 1912, no distinction was drawn between astronomy and astrology. If the Chinese 'astrologer' on the Turon was actually an astronomer using xingzhan, what knowledge and tools would he have required to predict rain?[10]

As long ago as the second century BCE, the ancient Chinese text, the Huainanzi 淮南子, compiled by the King of Huainan of the Han Empire, listed two methods of forecasting weather for military actions. One was houxing 候星, which originally referred to officials known as "Watchers of the Stars". Hence the technique of houxing is to observe the night sky. The other, wangqi 望氣, refers to the observation of meteorological phenomena, including

---

[8] "Bathurst", *The Freeman's Journal*, 11 November 1865, https://trove.nla.gov.au/newspaper/article/115450831

[9] "A Visit to Joss", *Evening News*, 8 March 1902, https://trove.nla.gov.au/newspaper/article/113867884

[10] The pronoun *he* is used because census records for the Sofala Registry District in 1861 and 1871 do not record any women born in China.

those related to the celestial objects such as meteors, meteorites and comets, as well as the sun and the Moon. As historian of Chinese science and technology Joseph Needham pointed out, the Chinese were long ahead of the West in certain methods of meteorological measurements and kept records of a more complete nature over a much longer time.[11] Some of the meteorological phenomena were compiled in household encyclopedias, riyong leishu 日用 類書, in common use from the time of the Ming dynasty in the 15th century as a quick daily reference to forecast weather. The London Missionary Society (LMS) collection in the National Library of Australia has one such household encyclopedia, Zengbu Wanbao quanshu 增補萬寶全書, which contains the divinations of meteorological phenomena in the first chapter.[12] Assuming the rain forecast on the Turon was made using Chinese astrology rather than non-astrological divination techniques, the astrologer on the Turon would have used either wangqi or houxing to make his prediction.

For Watchers of the Stars, rain could be predicted by the astronomical event of the Moon approaching the constellation Bisu 畢宿 (the net constellation, i.e. Hyades). Bisu is one of the twenty-eight lunar mansions, ershíba xiu 二十八宿, a constellations system reflecting the movement of the Moon through a sidereal month. A poem on this astronomical event appeared in the Shijing 詩經, an anthology of Chinese poetry written between the 11th and 7th centuries BCE, compiled by Confucius and regarded as one of the "Five Classics" of Confucianism.

> 有豕白蹢、烝涉波矣。
> 月離于畢、俾滂沱矣。
> 武人東征、不遑他矣。
> There are swine, with their legs white,
> All wading through streams.

[11] Joseph Needham, *Science and Civilisation in China: Volume 3, Mathematics and the Sciences of the Heavens and the Earth*, (Cambridge: Cambridge University Press, 1959), p. 926.

[12] *Zeng bu Wan bao quan shu / zhu ming jia hui ji ; Mao Huanwen zeng bu shi*, National Library of Australia, LMS 254, digitised item, https://nla.gov.au/nla.obj-46009837/view?partId=nla.obj-46009942. Last accessed 24 September 2019.

The Moon also is in the H[y]ades,
Which will bring still greater rain.
The warrior, in charge of the expedition to the east,
Has no leisure [to think] of anything but this.[13]

In Chinese mythology, Hyades represents the deity of rain, Yushi 雨師. In ancient times, the Chinese worshipped Yushi and offered a rain sacrifice, Yusi 雩祀, to plead for rain in times of drought. While this poem associates the Moon in Hyades with rain, it does not explain the occurrence of rain, as it does not rain every time the Moon approaches Hyades. More precise conditions were offered in 80 BCE by influential philosopher Wang Chong in the Lunheng 論衡, which discussed natural science and criticised Chinese philosophy and metaphysics. One chapter criticised the rain sacrifice. It referred to one of the Confucian analects to clarify the circumstances which determine the occurrence of rain:

> Confucius, on the point of going out, bade Tse Lu [a student of Confucius] prepare his rain apparel, and, after a few minutes, in fact, a great shower came down. Tse Lu asked for an explanation, and Confucius replied, "Yesterday evening the Moon approached the Hyades." Later on, the Moon had again approached the Hyades. Confucius going out, Tse Lu wished to prepare his rain apparel, but Confucius would not have it, and really it did not rain, after he had left. Tse Lu asked the reason. "Formerly," said Confucius, "the Moon drew near the northern part, hence it rained. Yesterday evening the Moon came near the southern part, therefore it did not rain."[14]

Further Wang wrote:

> When the Moon proceeds on the northern way and approaches the northern part of the Hyades, it nearly always rains. Accordingly, the Hyades must be situated on the northern way. But would this constellation of the northern

---

[13] Jean Elizabeth Ward (ed.), *Book of Odes English Edition*, (lulu.com, 2008). p. 150.
[14] Alfred Forke, *Lun-Heng*, Part II, (New York: Paragon Book Gallery, 1962). pp. 328-329.

way be willing to send down rain, in response to a rain sacrifice? When Confucius was going out and calling upon Tse Lu to get his rain apparel ready, there certainly was no rain sacrifice offered in Lu [a vassal state during the Zhou dynasty] simultaneously, and, notwithstanding, torrents of rain came down spontaneously, and without any prayer there was bright sunshine again of itself. Thus fine weather and rain have their times. In the course of a year, sunshine and rain alternate. When there is to be rain, who must pray for it, and when there is to be sunshine, who can stop it?[15]

Although Wang criticised the rain sacrifice he still believed that under certain conditions, the Moon in Hyades would cause rain. By contrast, when the Royal Astrologer, Chinese-born Indian, Gautama Siddha, compiled the Great Tang Treatise on Astrology of the Kaiyuan Era (Kaiyuan Zhan Jing 開元占經) in the Tong Dynasty around 720, his chapter on rain divination made no mention of it.

In the eight hundred years following the publication of Wang's treatise, precession shifted the positions of the constellations, affecting the accuracy of the prediction. Watchers of the Stars with astronomical and calendrical knowledge were able to make adjustments for precession. They kept their knowledge a trade secret with the exception of Song Dynasty polymath scientist Shen Kuo 沉括 (1031-1095) who shared this knowledge in his book Mengxi Bitan 夢溪筆談. The next adjustment was made in the Ming Dynasty. The government official Xu Guangqi 徐光啟, influenced by Western astronomy introduced by the Jesuits, submitted the "Memorial on a List of Proposals to Correct the Precessions in Calendar Making" ("Tiaoyi Lifa Xiuzheng Suicha Shu" 條議曆法修正歲差疏) in 1629 to Chongzhen, the last Emperor of the Ming Dynasty. Later that year the calendrical reformation began. The first draft calendar, completed in 1634, was used by the Manchurian court throughout the Qing Dynasty (1644-1912).

Astronomical and calendrical knowledge was compiled by court officials in the Imperial Almanac, while tongshu 通書, compiled with additional astrological and astronomical information, were

---

[15] Ibid, p. 331.

reprinted for the commercial market. Like farmers almanacs published in America, tongshu provided chronomancy information on the best days for farming, fishing, and hunting. Copies of tongshu found in the collections of Australian libraries indicate that the Chinese brought tongshu with them to Australia.

Between 1848 and 1853, an estimated 3,000 Chinese indentured labourers were brought to New South Wales to work in the pastoral industry, most of them on five-year contracts. Arriving without possessions, they were sent to work beyond the frontiers of settlement in the Central West, New England, Murrumbidgee, the Darling Downs, and the Port Philip districts. The majority of these labourers left from the port of Amoy in Fujian province. Fujianese were mostly fishermen and seafarers, who relied on astral navigation, namely guoyang qianxing 過洋牽星. The indentured labourers from Fujian would have been disoriented under the unfamiliar southern sky. It has not yet been ascertained if any of the first wave of Chinese migrants to the Australian colonies brought tongshu with them.

The second wave of Chinese migrants were generally peasant-farmers-cum-gold-seekers who began arriving from 1854, most of them from the Pearl River Delta. They cared less about the night sky and more about the seasons. They would have used the 24 solar terms (ershisi jieqi 二十四節氣) of the Chinese lunisolar calendars to stay synchronized with the seasons, and the 72 seasonal indicators (qishier hou 七十二候), which sub-divide the solar year into 72 terms as listed in tongshu.[16] These gold-seekers were led to the goldfields by headmen, often storekeepers, who brought with them tongshu almanacs, which were not always helpful in the southern hemisphere.

In 1856, the Scottish-Malaysian Chinese missionary William Young reported a conversation he had with a Chinese storekeeper in Castlemaine. The storekeeper commented to Young that neither the lunar eclipse that year, nor the solar eclipse of 5 April 1856 were anticipated in the Chinese calendar for the present year, which, he said, rendered members of the Astronomical Board in Peking liable to the punishment of having their heads cut off

---

[16] Jean-Claude Martzloff, *Astronomy and Calendars – The Other Chinese Mathematics*. (Verlag Berlin Heidelberg: Springer, 2016), p. 67.

in consequence of their gross oversight. Young explained to the storekeeper that an eclipse visible here would not be visible in China and that the Chinese calendar would mention only those eclipses visible in China and not elsewhere.[17] The storekeeper asked Young what occasioned eclipses, what was lightning, and where rain and clouds came from.[18]

Newspaper reports of the reactions of these Chinese gold seekers to the lunar eclipse on the 20 April 1856 show they acted according to superstitious beliefs.

> [The eclipse] seemed to excite more attention among Chinese than any other portion of our inhabitants … in Chinese encampments from Red Hill to Little Bendigo "gongs, pots pans, tin dishes, and everything that when beat would give forth a sound were called into requisition.[19]

Similarly, the *Bendigo Mercury* reported that the lunar eclipse of 13 and 14 August 1859 was met by the "discordant music of the Chinese gongs."[20] Decades later Australian adventurer George.E. Morrison who was in Tongchuan in Yunnan province, China, at the time of the solar eclipse of 6 April 1894 wrote:

> the town was in commotion; kettledrums and tom-toms were beating, and crackers and guns firing; the din and clatter was continuous and deafening.

The reason for this was explained by the belief that the sun was being swallowed by the Dog of Heaven, and the noise was to compel the monster to disgorge its prey.[21] The banging of drums was an official ritual organised by local courts around China to mobilise the people to protect the Emperor. The sun represented

---

[17] "Chinese Mission", *The Mount Alexander Mail*, 13 May 1856, https://trove.nla.gov.au/newspaper/article/202633749

[18] Ibid.

[19] *The Ballarat Star* quoted in "The Royal Charter", *The Mount Alexander Mail*, 2 May 1856, p. 3.

[20] "The Eclipse of the Moon", *Launceston Examiner*, 1 September 1859, https://trove.nla.gov.au/newspaper/article/38997210

[21] Morrison, George. *An Australian in China: being the narrative of a quiet journey across China to Burma*. Third Edition. (London: Horace Cox, 1902), p. 125.

the Emperor, and the solar eclipse was believed to foreshadow disaster for the Emperor. Astronomical phenomena, especially infrequent and extraordinary occurrences, were considered omens or portents, which could influence the lives of China's rulers and its people. The Moon represented the people, and the sight of a red Moon during a lunar eclipse was believed to foreshadow famine or disease. Such phenomena, therefore, needed to be foretold in the Imperial Almanac.

Young's conversation with the Castlemaine storekeeper revealed a problem with using the almanacs in Australia. That is, the astronomical data related to the Northern Hemisphere. To be useful in Australia, the data needed to be adjusted. Given the ignorance of astronomical and meteorological phenomena shown by the Castlemaine storekeeper and the superstitions of the gold seekers, is it possible that there was an astronomer with the skill to adjust astronomical data in the Almanac for the Southern Hemisphere?

While there is no other mention of 'astrologers' on the goldfields, in 1859, the correspondent for the *Sydney Morning Herald* commented on the level of literacy and numeracy amongst the Chinese community at Golden Point on the junction of the Meroo and Grattai Creeks:

> Chinese butchers, storekeepers and tavern-keepers under-stand their respective trades quite as well as the same class of Europeans on the goldfields, and are, as a community, much better educated. They can all read, write and keep correct accounts, while it is no uncommon thing to find a licensed publican or a storekeeper amongst the Europeans who can neither write nor read.[22]

There was certainly a need for accurate rain prediction on the goldfields. Unlike Canton (present-day Guangzhou), where the monsoon brought seasonal rain, rainfall on the leeward side of the Great Dividing Range was less predictable and more subject to drought. The success of the gold-seekers was dependent on the availability of water to wash the pay-dirt and extract the gold. As

---

[22] "Visits to the Western Goldfields", *The Sydney Morning Herald*, 20 May 1859, https://trove.nla.gov.au/newspaper/article/13025220

Matthew Higgins, historian of the Turon goldfields put it:

> Too little water brought mining operations to a standstill
> … too much water—in the form of floods—could equally
> spell disaster … mining methods and equipment on the
> Turon were shaped as much by water as they were by
> gold.[23]

When *Sydney Morning Herald* correspondent, Charles De Boos, visited the new goldfield of Glanmire in September 1865, the field excluded Chinese, but after his visit, he learned that the field had opened to Chinese miners.[24] De Boos wrote that Glanmire was "hardly the kind of place that will suit John Chinaman, especially with £1 per month to pay, and not more than about a hundred of them have availed themselves of the opportunity to occupy."[25] The Bathurst district was in drought for most of 1865, and, by October, the almost certain failure of the wheat crop was predicted, grain and hay were being purchased at famine prices and bushfires swirled around the district.[26] At Glanmire, the creek was all but dry, and the European miners, hamstrung by the lack of water needed for washing gold, began to desert the field.[27] Yet, in October 1865, Chinese began taking out an increasing number of licences. The *Herald* reported "a number of Chinese have settled on the field and the sight presented on the flat, at the head of the township, is one

[23] Matthew Higgins, *Gold and water: A history of Sofala and the Turon goldfield*, (Bathurst: Robstar Pty Ltd, 1990), p.4.
[24] "Bathurst", *The Freeman's Journal*, 11 November 1865, https://trove.nla.gov.au/newspaper/article/115450831; Glanmire opened to Chinese miners on 19 September 1865, see "Telegraphic Intelligence", *The Maitland Mercury*, 9 September 1865, https://trove.nla.gov.au/newspaper/article/18702439
[25] "Random Notes No. XXVI", *The Sydney Morning Herald*, 7 October 1865, http://trove.nla.gov.au/newspaper/article/13119988
[26] "New South Wales", *The Perth Gazette and West Australian Times*, https://trove.nla.gov.au/newspaper/article/3756714; "Country News", *Freemans Journal*, 14 October 1865, https://trove.nla.gov.au/newspaper/article/115452981
[27] "The Glanmire", *The Sydney Morning Herald*, 7 November 1865, https://trove.nla.gov.au/newspaper/article/31126149

of busy industry."[28]

What perhaps encouraged the Chinese to move onto the field when there was no end to the drought in sight was the prediction made by the Chinese astrologer that rain would fall on the first Friday in November, which indeed it did.

Wing-Fai Wong is an information technology professional with a strong interest in Chinese metaphysics. While working in IT for a living, he continues to undertake research on Chinese metaphysics, including fengshui (geomancy) and zeri (chronomancy), in the everyday life of Chinese Australasians. His conference presentations include "The Significance of Luban Jing" for Dragon Tails 2015, "Five Emperors Coins" for the 2017 Symposium of the *Academic Journal of Fengshui – Oceania*, and "Chinese Merchants and Fengshui" for the 2019 International Conference on Chinese Entrepreneurship and Social, Economic, and Political Transformations of China and the World.

Juanita Kwok was born in Sydney and gained an arts degree at the University of Sydney before moving to live in Bathurst in 2008. In 2013 she wrote her Honours thesis at Charles Sturt University on the representation of Chinese in Australian feature films made in the White Australia era. She was awarded a scholarship as a PhD student, and has just completed her thesis, "The Chinese in Bathurst: Recovering Forgotten Histories". She looks forward to graduating in December 2019.

---

[28] "The Glanmire", *The Sydney Morning Herald*, 31 October 1865, https://trove.nla.gov.au/newspaper/article/31125871